T0136418

User Requirements for Wireless

WIRELESS WORLD
RESEARCH FORUM

RIVER PUBLISHERS SERIES IN COMMUNICATIONS

Volume 42

Series Editors

ABBAS JAMALIPOUR
The University of Sydney
Australia

MARINA RUGGIERI
University of Rome Tor Vergata
Italy

HOMAYOUN NIKOOKAR
Delft University of Technology
The Netherlands

The "River Publishers Series in Communications" is a series of comprehensive academic and professional books which focus on communication and network systems. The series focuses on topics ranging from the theory and use of systems involving all terminals, computers, and information processors; wired and wireless networks; and network layouts, protocols, architectures, and implementations. Furthermore, developments toward new market demands in systems, products, and technologies such as personal communications services, multimedia systems, enterprise networks, and optical communications systems are also covered.

Books published in the series include research monographs, edited volumes, handbooks and textbooks. The books provide professionals, researchers, educators, and advanced students in the field with an invaluable insight into the latest research and developments.

Topics covered in the series include, but are by no means restricted to the following:

- Wireless Communications
- Networks
- Security
- Antennas & Propagation
- Microwaves
- Software Defined Radio

For a list of other books in this series, visit www.riverpublishers.com
http://riverpublishers.com/series.php?msg=Communications

User Requirements for Wireless

Editors

Lene Tolstrup Sørensen

Associate Professor
Aalborg University
Denmark

Knud Erik Skouby

Professor, Director CMI
Aalborg University &
Chair WGA/WWRF
Denmark

WWRF Book Series Editor:
Knud Erik Skouby

River Publishers

Published, sold and distributed by:
River Publishers
Niels Jernes Vej 10
9220 Aalborg Ø
Denmark

ISBN: 978-87-93237-20-9 (Hardback)
        978-87-93237-61-2 (Ebook)
©2015 River Publishers

Foreword

The WWRF Series in Mobile Telecommunications

The Wireless World Research Forum (WWRF) is a global organization bringing together researchers into a wide range of aspects of mobile and wireless communications, from industry and academia, to identify the key research challenges and opportunities. Members and meeting participants work together to present their research and develop white papers and other publications on the way to the Wireless World. Much more information on the Forum, and details of its publication programme, are available on the WWRF website www.wwrf.ch. The scope of WWRF includes not just the study of novel radio technologies and the development of the core network, but also the way in which applications and services are developed, and the investigation of how to meet user needs and requirements.

WWRF's publication programme includes use of social media, online publication via our website and special issues of well respected journals. In addition, where we have identified significant deserving subjects, WWRF is keen to support the publication of extended expositions of our material in book form, either singly-authored or bringing together contributions from a number of authors. This series, published by River Publications, is focused on treating important concepts in some depth and bringing them to a wide readership in a timely way. Some will be based on extending existing white papers, while others are based on the output from WWRF-sponsored events or from proposals from individual members.

I hope that each volume of this series will be useful and informative to its readership, and will also contribute to further debate and contributions to WWRF and more widely.

Dr Nigel Jefferies
WWRF Chairman

Contents

5 Security and Usability   77

Jing Chen, Marcus Wong and Lijia Zhang

Preface

This book is inspired by ongoing work and activities in World Research Wireless Forum (WWRF). With the growing trend of wireless devices, tags and sensors, there is also a growing interest in development of software, apps for controlling daily life, news or gaming. The number of apps is continuously growing promoted by the activity of an increasing number of software developers in the world. The discipline of software engineering is thus becoming more and more important and this book takes a User-Centered Design approach to software engineering focusing on user centric projects and cases. The projects and cases represent different socio-economic settings and sectors providing insight to different types of user requirements by changing of the user group in age, the context and the different technologies or software. By focusing on different case studies, the book provides insight into practical user requirements elicitation which all describes the process of software development with some level of user involvement.

The book is addressing software engineer and university audiences.

Sincerely thanks to authors dr. **Kari Heikkinen**, Associate Professor, the Lappeenranta University of Technology; dr. **Pirkko Paananen-Vitikka**, Adjunct Professor, University of Oulu; dr. **Jari Porras, Professor** at the Lappeenranta University of Technology; **Ross Purves**, Lecturer at De Montfort University; **Graham Welch, Professor** at University of London; dr. **Philip Edge**, Associate Consultant at eNovation4D; dr. **Harsha Liyanage**, Lead Consultant eNovation4D; **Jannick Kirk Sørensen**, Assistant professor at Aalborg University; **Jing Chen**, senior researcher, Huawei's Shanghai Research Center; **Marcus Wong**, Huawei Technologies (USA); **Lijia Zhang**, Huawei's Research Center in Beijing.

We are grateful for inspiration and support from the publishers, ITU and WWRF.

Finally, but not least thanks to Lene Tolstrup Sørensen getting the idea of the book and for her tireless effort in collecting the contributions and editing.

Knud Erik Skouby
Copenhagen, January 2014

List of Abbreviations

3G	Third Generation of Mobile Telecommunications Technology
ADHD	Attention Deficit Hyperactivity Disorder
AuC	Authentication Center
BBC	British Broadcast Corporation
DR	Danish Broadcast Corporation
DOS	Denial of Service
GPS	Global Positioning System
ICT	Information and Communication Technologies
ICT4D	ICT for Development
IMS	IP Multimedia Subsystem
JamMo	Jamming Mobile
LTE	Long Term Evolution (4G)
M4D	Mobile for Development
NFC	Near Field Communication
NGO	Non-Governmental Organisation
PDA	Personal Digital Assistant
PSB	Public Service Broadcaster
RFID	Radio Frequency Identification
RSS	Rich Site Summary
SMS	Short Message Service
SIM	Subscriber Identity Module
UMSIC	Usability of Music for the Social Inclusion of Children (EU Project)
UMTS	Universal Mobile Telecommunications System

List of Figures

List of Tables

Introduction

Sørensen, L. and Skouby, K. E.

Center for Media and Information Technologies,
Aalborg University, Aalborg, Denmark

The World Wireless Research Forum has for years had a vision forecasting: "7 trillion wireless devices serving 7 billion people by 2017" (Jefferies, 2008). This would correspond to around a thousand devices per user including RFID tags, sensors, and all other technologies that users use in daily life. Following the technology trends, we are getting closer to the vision and most people experience a sense of mobility and use of wireless technologies they did not have any idea of 10 years ago. Taking mobile adoption as one example of the fast increasing wireless use, in 2014 32% of the world population owned a smartphone (ITU, 2014)—in 2010 the number was 9%. The ITU numbers (ITU, 2014) imply that now, there are more mobile devices on Earth than people, which shows a trend toward the WWRF vision. The trend is not only in developed countries. The book by Skouby and Williams (2014) presents different situations of the use of mobiles in Africa. It is here said that in 2013, more than 6% of the region's gross domestic product (GDP) originated from the mobile industry (Skouby and Williams, 2014), making it an important part of Africa's economy.

With the growing trend of wireless devices, tags, and sensors, there is also a growing interest in development of software, apps for controlling daily life, news, or gaming. The number of apps across different stores corresponds to almost 9 million (PC Magazine, 2014). Estimates say that there are 18.5 million software developers in the world (Ranger, 2013), which places focus on the importance of software development in today's societies. The discipline of software engineering is becoming more and more important with the increasing use of mobile devices.

Software engineering is a traditional discipline used for development of software (Sommerville, 2011). Sommerville (2011) identifies the following

activities of software engineering: software specification, development, validation, and evolution. Different types of systems needs different development
processes, and therefore, emphasis can be put one or the other of the software
engineering activities.

Focus of this book is on the user requirements of software engineering.
Following Sommerville's (2011) terminology, this is a part of the specification
that often initiates any software development process. User requirements are
generally defined "as statements of what services the system is expected
to provide to system users and the constraints under which it operates"
(Sommerville, 2011). It covers the functionalities and needs that the user
might have to a particular system or software.

In any textbook on IT system development, the identification or elicitation
of user requirements is a key building block (see, for example, Lauesen, 2002;
Sommerville, 2011). In practice, user needs and wants are a challenge and
faulty identifications in this part of an IT system development can cause huge
problems with the final product (Stålbröst and Bergvall-Kåreborn, 2008).
Hickey and Davis (2004) furthermore describes requirements elicitation as
one of the most critical activities of software development causing failures
of projects if not taken well care off. In this book, requirement elicitation is
preferred to identifying, gathering, or capturing of requirements. Elicitation
signals that requirements not just can be collected. They have to be interpreted,
analyzed, modeled, and validated before they can be used for software
development (Nuseibeh and Easterbrook, 2000).

Discussions are often seen on whether to involve users in the process
for user requirements (Edvardsson et al, 2000). However, if focus is on the
elicitation process, it is clear that the user involvement is an essential and
necessary part of the process. Therefore, this book takes a user-centered
design (UCD) approach to software engineering focusing on user-centric
projects and cases. User-centered design is defined by a number of elements
(Oulasvirta, 2005):

- The role of the human needs is central and directing innovation and design
- Focus on understanding the users in their natural use contexts
- Enhancing people's activities and tasks through technology.

User-centered design clearly has a focus on involvement of users in the
project. User involvement can take place in a number of different ways:
through interviews, focus groups, to prototyping, co-creation (Battarbee and
Koskinen, 2005), and living labs (Ståhlbröst, 2008)—increasing the user
involvement level. However, the elicitation of user requirements also can take

place by use of audio-visual information, sharing of user stories, and setting up repositories of these to support a larger group of users (see, for example, the StoryBank Project; Frohlich et al, 2009a, b). User involvement is a perspective that has been put forward by the agile manifesto (Agile Alliance, 2001) that describes user involvement as imperative to any software development process.

User requirements as such change according to age groups, gender, and experience in use of the system/software, while in some situations, they change over time (Kelly, 2004). User requirements, therefore, cannot be exchanged directly between projects, IT systems, and different software. That makes the elicitation of user requirements an inherent part of any software development project and a resourceful activity as well.

This book aims at providing insight to different types of user requirements by describing varying case studies in which technologies or software have been developed. It has been sought to provide a variety of user require-ments by changing the user group in age, the context, and the different technologies or software. The book is foreseen to provide insight into practical user requirements elicitation by focusing on different case studies, which all describe the process of software development with some level of user involvement. The contexts are changed to address the challenges in practical user requirement elicitation. The book has representation from 4 different contexts: a school, a construction site, farmers in Sri Lanka, and a Web environment for personalized media services. Furthermore, there is a chapter on usable security requirements, which emphasizes the need to think security more into any sort of service being made. The book is outlined as follows:

In *Chapter 1*, focus is on the design of a mobile application for social inclusion of children in a targeted group of children with attention deficiencies and children of immigrants, having a language different from the country they live in. The case study is based on a EU FP7 project named UMSIC (www.umsic.org). The mobile application was developed to support children from ages 3 to 12 years with three different entries according to age group. The software development process followed the ISO13470 standard for user involvement where the children were involved in parallel to software design to elicit requirements. The project produced 98 use cases from which user requirements were derived to support the application development. The chap-ter describes the application, how the application was developed and discusses the impact the user involvement had on the development process. Furthermore, there are discussions on the challenges in the project and guidelines for future software development.

Chapter 2 describes how a mobile application has been designed to support the construction industry in planning the construction by a tight communication between involved partners. This is seen as key for a successful project outcome and for reducing the costs and resources spent on any construction project. The chapter describes the special conditions the construction industry is facing in its work. These elements serve as requirements for the application development. The chapter describes the user studies made, the development of the application, and the lessons learned. During the development of the application, lessons were learnt concerning the device, the project start-up, and about some limitations for input and modality feedback from the application.

The development of a mobile phone application for rural farmers in Sri Lanka is described in *Chapter 3*. The chapter describes how a local NGO, Fusion, has founded the application development, named FarmerNet, in support of the rural farmers, so they can interact directly with traders and thereby controlling their own income in a better way. Much focus has been put into the involvement of the farmers in the innovation process that describes the application development. The process is described through elements such as idea creation, concept development, and prototype development. The chapter describes criteria to judge the appropriateness of a technology and how the FarmerNet application has secured this in the development process.

Chapter 4 reflects on the introduction of customized and personalized news introduced by the Danish Broadcasting, DR, on their news web page. The chapter discusses the concepts of user requirements and questions whether this is a valid concept in particular when other stakeholders, as the leaders of a news section on a broadcasting Web site expresses the user requirements. A case study, My DR, is being presented describing how the customization was done in 2005 and what came out of this. It turned out that the customization of the web page not was as popular as was foreseen from the service provider's side. The conclusions of the chapter are that user requirements are not easy to construct in particular when different stakeholders are involved in the requirement process.

Chapter 5 places focus on one element crucial to most application development, namely security. More specifically, the chapter provides requirements for what is called usable security. Usable security is about making security transparent and understandable to the extent possible at the same time as making it visible to give the user control and understanding of the security elements existing in any IT-based interaction. The chapter analyses current mobile communication systems in respect of security status as well as the stakeholders involved in mobile communication systems. This is used to suggest a number of usable security elements that should be taken into consideration for design of future mobile systems and software.

Finally, *Chapter 6* provides a summary of the learning of the different case studies presented in the book. The conclusions furthermore include a section on guidelines derived from the case studies of this book.

References

[1] Agile Alliance (2001): The Agile Manifesto, http://www.agilealliance.org.

[2] Battarbee, K., and I. Koskinen (2005). "Co-experience: User Experience as Interaction," *CoDesign* 1, no. 1 (2005): 5–18.

[3] Edvardsson, B., P. Gustafsson, P. Magnusson, and J. Matthing, eds. (2000): *Involving Customers in New Service Development. Series on Technology Management*, Vol. 11. Sussex: Imperial College Press.

[4] Frohlich, D., R. Bhat, and M. Jones (2009): "Democracy, Design and Development in Community Content Creation: Lessons From the Story-Bank Project." *Annenberg School for Communication & Journalism* 5, no. 4 (2009a): 19–35.

[5] Frohlich, D., D. Rachovides, K. Riga, R. Bhat, M. Frank, E. Edirisinghre, D. Wickramanayaka, M. Jones, and W. Harwood (2009): *StoryBank: Mobile Digital Storytelling in a Development Context*. CHI 2009–Mobile Applications for the Developing World. April 8th, 2009, Boston, MA, USA (2009b).

[6] Hickey, A.M., and A.M. Davis (2004): "A Unified Model of Requirements Elicitation." *Journal of Management Information System* 20, no. 4: 65–84.

[7] ITU (2014), Key ICT indicators for developed and developing countries and the world (totals and penetration rates). ITU_Key_2005-2014_ICT_data.xls. Accessed 23.03.2015

[8] Kelly, A. "Why do Requirements Change?" (2004). http://accu.org/index .php/journals/319

[9] Lauesen, S. (2002): *Software Requirements. Styles and Techniques*, 608. Addison-Wesley: Pearson. ISBN 0 201 74570 4.

[10] Nuseibeh, B., and S. Easterbrook (2000): "Requirements Engineering: A Roadmap." *ACM 2000*, 1-58113-253-0/00/6. (2000)

[11] Oulasvirta, A. (2005): "Grounding the Innovation of Future Technologies." *Human Technology* 1, no. 1: 58–75.

[12] PC Magazine. (2014): Mobile Phone App Store Statistics. www.statistic brain.com/mobile-phone-app-store-statistics.

[13] Ranger, S. (2013): There are 18.5 million software developers in the world – but which country has the most? www.techrepublic.com/blog (european-technology/there-are-185-million-software-developers-in-the-world-but-which-country-has-the-most/.

[14] Sommerville, I. (2011): *Software Engineering*. Boston, MA: Pearson.

[15] Skouby, K.E. and I. Williams (2014): *The African Mobile Story*. Aalborg: River Publications, 2014.

[16] StoryBank: Digital World Research Centre, http://www.dwrc.surrey.ac .uk/storybank.shtml.accessed 23.03.2015

[17] Ståhlbröst, A. (2008): Forming Future IT. The Living Lab Way of User Involvement. Ph.D. thesis at Luleå University of Technology, Social Informatics, Sweden.

[18] Stålbröst, A. and B. Bergvall-Kåreborn (2008): "Constructing Representations of Users Needs–A Living Lab Approach." Asproth, V., *IRIS31—Public Systems in the Future; Possibilities, Challenges and Pitfalls*, 10–13 August, at Åre, Sweden.

1

Designing Mobile Applications for Children

K. Heikkinen[1], T. Kallonen[1], P. Paananen[2], J. Porras[1], R. Purves[3], J. C. Read[4], T. Rinta[4] and G. Welch[4]

[1]Communications Software Laboratory, Lappeenranta
University of Technology, Lappeenranta, Finland
[2]Department of Music, University of Jyväskylä,
Jyväskylän yliopisto, Finland
[3]Department of Culture, Communications and Media,
Institute of Education, University of London, London, UK
[4]University of Central Lancashire, Preston, Lancashire, UK

1.1 Introduction

As children become primary users of technologies, there is a need to identify good practice in designing for them. Previously, before interactive technology was so mainstream, most applications for children were educational and were designed primarily with educational goals in mind as opposed to them being designed primarily with the children in mind. Early products of this type tended to feature bright colors, animal like characters and simple writing, but that, to a significant degree, was the extent of the "design for children." As computer technology has moved from the classroom into children's homes, and more recently into children's hands, there has been an increase in interest in discovering how best to design for children. At the same time, the use of technology by children has changed with much more technology being designed for fun, for docility, and for play; education is no longer the main interest in this arena.

Designing for children requires a multifaceted approach; designers need to understand the needs of children as well as the needs for the software. Thus, where an application is needed to provide increased sociality, the designers need to understand social behavior. In addition, where specific technologies are used, designers have to be aware of the limitations of, and possibilities for, the chosen hardware. This makes designing for children an interesting and challenging activity. Designing for children has been studied by the Interaction Design and Children academic community both in terms of designing for and designing with children. In the former, the interest is mainly about how children's software products can be created—many papers provide guidelines for design, for example Inkpen (2001), Gilutz and Nielsen (2002), and Brederode et al. (2005). In the latter, the focus is on how children can be brought into the process of design by encouraging their active participation (Scaife et al., 1997). Here, much of the discussion is about the extent to which children can participate in design, as found in Read et al. (2002).

Designing mobile applications for children is an area that has had a considerable attention; there has been a book on the subject (Druin, 2009), there have been papers discussing how mobile phones should be designed differently for children (Read, 2009), and there have been studies about how children use mobile technologies (Hart, 2007). In most of the work to date on mobile technologies, the emphasis has been on technologies that are primarily used as phones. While this is clearly a major use of mobile phone technology, the secondary uses, for music, for games, and for clock and calendar functions, as alluded to in Read (2009), is clearly significant.

Phones that can be used as other devices offer enticing possibilities for novel design. Because they are embedded with the hardware and software to give them connectivity, because they are generally able to offer good screen resolution and easy interaction, these devices are especially interesting for exploitation in social, connected settings. Designing for these uses is challenging; the UMSIC project as the background for this chapter, took on this challenge.

1.2 JamMo, the Product of UMSIC Project

The UMSIC—Usability of Music for the Social Inclusion of Children project (www.umsic.org)—was a multidisciplinary EU FP7–funded project aimed for improving social inclusion and reducing isolation in groups of children, especially targeting children with attention deficiencies and children of immigrants, whose language is different from that of the host country. The goal was to provide the children with a tool they could use together to create something new in groups and by doing so including everyone in the process and thereby enhancing social inclusion in general. Table 1.1 summarizes the project objectives that are considered while focusing on designing for the children.

Table 1.1 Project objectives

Inclusion Objectives
Learning disabilities are a major threat for positive child development. Among the most common types of learning disabilities are attention deficit disorders. These children are at risk of impaired functioning in most areas related to school success. Another challenge for today's societies is the increasing globalization, and as a result, the enormous international migration due to economic reasons and due to political and social conflicts. European cities nowadays are largely international and, as a consequence, children grow up in multicultural contexts
Education Objectives
In order to offer children better chances to integrate into the multicultural European society, the UMSIC project assumes that language and music are essential for building a cultural identity and for actively participating in various cultural environments. In addition, language and music are both powerful means to integrate into new social groups
Technological Objectives
The UMSIC project aims at developing a mobile time- and place-independent application that provides an interactive tool for children while also enabling them to communicate musically and informally with their peers. The objectives were met through three related components, namely (1) a technical solution, (2) its application, and (3) target users. The technical component specifies the different operation modes of the technical environment, that is, stand-alone, networked, ad hoc, and public. The application component specifies the range of possible built-in features, that is, improvisation, composition, karaoke, and virtual instruments. The third component, target users, specifies the special characters of children

A central part of our approach was a mobile application, named JamMo (Jamming Mobile) presented in Figure 1.1 (Welcome screen) and 1.2 (Main screen). JamMo is an interactive product that enables children to communicate informally with their peers by using mobile technologies (Nokia N900 was selected as the target platform for the application). The JamMo shows a special focus on child-centered usability, intelligent musical engineering, and carefully developed pedagogical design that is allied to structured learning material.

JamMo is essentially a showcase that combines the use of technology and effective educational principles to foster social inclusion and prevent marginalization of those children in more challenging contexts. JamMo was designed to address the concerns of targeted groups of end-users, that is, preschool (aged 3–6) and school-aged (6–12) children with an emphasis on those with moderate learning difficulties and those who are immigrant and studying in unsupported integrated school classes. JamMo aims to support children's processes of social inclusion through the use of music technology and touch screen input device. JamMo allows children of different age groups

Figure 1.1 The door that opens the JamMo application (based on design by children)

Figure 1.2 The JamMo 3–6 main screen and the cupboard used to store the music created by the children

to create and share music using different methods in different usage scenarios. In these scenarios, depending on the nature of the scenario, children can sing, compose with predefined sound loops, or play with virtual instruments.

1.2.1 JamMo for 3–6 Years Old

JamMo for 3–6 years old consists of a singing game and a composition game. The singing game aims to encourage the children to sing along with a premade song. After singing, the child can listen his/her singing. The game contains songs with four languages or can be listened only with instrumental accompaniment (see Figure 1.3). The composition game offers three themes (City,

Figure 1.3 The song selection menu and the singing game

Figure 1.4 The theme selection menu and composition game with city theme

Animal World, and Fantasy; see Figure 1.4). Each theme has preset sound fragments and supporting backing track. The child can build compositions by dragging and dropping these sound fragments (icons on the UI) to a sequencer track in the bottom of the screen. The bear in the upper left side of the screen is the Mentor and assists the child.

1.2.2 JamMo for 7–12 Years Old

JamMo for 7–12 is slightly more complex version in its operation and functionality than JamMo 3–6. The sequencer has now four tracks instead of 1–2 tracks for younger children (see the left part of Figure 1.5). It also contains sound sample wheels for selecting the sound fragments used in compositions and the ability to change the pitch and tempo of the song. This version contains orientation games that support the learning of features of JamMo and virtual instruments (e.g., keyboard in the right side of Figure 1.5) for improvisations and compositions. This version has also support for community features, such as sharing compositions and a discussion forum.

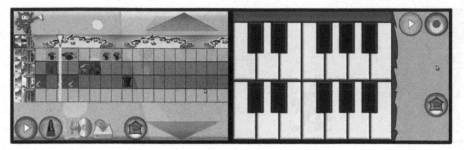

Figure 1.5   The four-track sequencer and the virtual keyboard

1.3 Toward Social Inclusion

Main activities to reach the objectives and expected impact can be roughly clustered under three domains, that is, design and requirements, open source software development, and enabling impact. It should be noted that the development teams often worked in parallel manner. Figure 1.6 illustrates the overview of activities in those domains. The upper section is roughly the first year of the project, and the lower section the subsequent two years. The

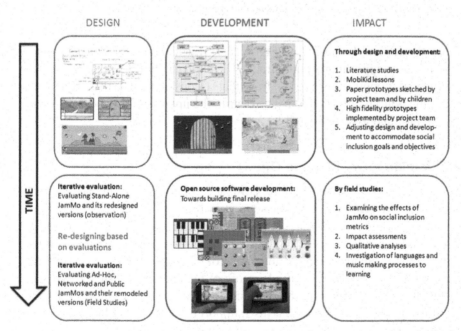

Figure 1.6   Overview of main activities

design was iterative and aided by sketches of project team and children. In the end, the designs were adjusted based on the results of the evaluations. The development followed open source software development guidelines and procedures. The aim of the activities in the end was to impact to raising social inclusion.

1.4 Design, Requirements, and Development

The project was aimed to be user-centric, in which the actual users (in our case, children) are the focus of the research and the development of the mobile application. Thus, it is essential to engage the users as integral part of both the research and development process. Most common standard in this area is ISO 13407 that suggests four essential activities: (i) requirements gathering, (ii) requirements specification, (iii) design, and (iv) evaluation. These activities should be carried out in an iterative manner until the objectives are met. In our case, the design and the development followed the cycle illustrated in Figure 1.7. The design and development started with understanding social inclusion, and in particular, how immigrant children and children with ADHD

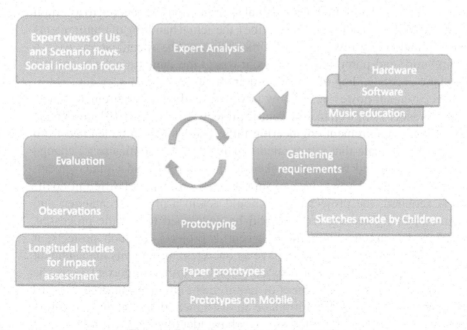

Figure 1.7 Design and development cycle

can become integrated through the mobile music software. This resulted an expert analysis of scenarios and flow of events on the user interface. This expert analysis triggered a requirement gathering for the JamMo. During that phase, children were also participating and drawing their own sketches to support the expert views. After requirements were gathered, the developers began to implement the first version of stand-alone JamMo. The development proceeded hand in hand with evaluation throughout the rest of the project. If new requirements were found, they would have been prioritized and implemented for final release if found important for the impact.

Requirements are an essential part of any (software) project. A requirement is a statement of being able to identify a factor or a certain characteristic that is important for the final product. Each requirement requires a distinct definition that is documented in an agreed manner by carrying out, e.g., elicitation, specification, analysis, and review of requirements. Requirements management, on the other hand, according to Hood et al. (2008) is a process that supports the requirements development, that is, producing those requirements. In designing JamMo, requirement elicitation was carried out in a distributed and parallel manner. Different teams, that is, interaction design, communications software, music engineering, usability-elicited requirements, provided them for steering board to approve. Even though the actual specifications were fixed to a certain timeline (to meet the deliverable deadline), the actual handling of requirements (analysis and review) was more flexible. Thus, as the specifications were a certain "mid" end-point in the process, it was at the same time the beginning of holistic understanding of various requirements.

It is essential, however, to know how well and what requirements engineering practices suit well to user-centric projects. Liu et al. (2010) present one survey where requirement engineering fails. The lessons to be learned from their survey indicate that requirements management activities (e.g., change management) should be tightly linked to the overall project management, appropriate domain knowledge and supporting prototypes (of different fidelities) is required, the project teams should act in a proactive manner, and requirements engineering tools should be developed to aid the processes. The same lessons are valid for user-centric projects. In such projects, it would be even more beneficial to have a concrete requirements management guideline with processes and responsibilities. Another important aspect is to be able to measure the success of requirement engineering processes. El Amam and Madhavji (1995) introduce an instrument that contains indicators of requirement engineering success. They have identified three dimensions of success: (i) *the cost-effectiveness*, (ii) *quality of the product*, and (iii) *quality of the requirement engineering service.* In UMSIC, a lot of resources were used in distributed manner, and no real widely accepted management process was used. This argument is debatable as, for example, comparisons to other projects requirement engineering has not been carried out. What we can also say is that changes in requirements were frequent (as the iterative flow of activities should produce). Also the fraction of the cost in requirement engineering was quite close to actual development. We believe that the quality of the product is good given the complexity of the project. However, real quality (toward inclusion) of the product can be only argued if impact is valuable. Quality of the product can be viewed also through how well the architecture is structured and cost-efficiency analysis. We can argue that from a modeling point of view, the rules were conformed and that good linking to objectives exists. However, some of the key issues, especially real-time issues, were not solved. The project did not have a structured cost efficiency analysis. The last dimensions subdimensions are how well the recommendations fit to the organization (which is not directly applicable to our project) and user satisfaction/commitment. The latter subdimension can be partially applied to our project as certain criteria (e.g., the degree of match between functionalities and user expectations) fit to the evaluation itinerary.

As the requirements aim to answer to questions such as what must be done and how we can verify we are doing the correct solution to our problem, the design should support the answering of those questions. Use cases are commonly used graphic visualization of desirable behavior of the solution. Use cases also assist us to find the individual objects and their hierarchies with associations. The UMSIC project produced 98 use cases. After use cases were derived from requirements, the flow of operation was built and graphically visualized by using UML sequence diagrams. Obviously, this required the information about the software architecture and its components and the communication between them. The software architecture was modularized to ease the development.

1.5 Impact

As noted in the earlier sections, the JamMo product is the centerpiece of enabling the expected impact. Consequently, a significant and on-going task was to evaluate the use of this software in a range of educational, social, and musical settings and with different users, particularly those at risk of being socially excluded (such as newly immigrant, or with moderate learning difficulties—see Section 1.1). An early evaluation example aimed to determine any special requirements of child users to ensure their participation and our usability evaluations. It was also important to be sensitive to the diversity of possible users and locations in our adopted evaluation research methods. Later evaluation, for example, embraced classroom-based, home-based, focus group and case study research studies.

One component of the impact research was a longitudinal study across one school year in the relatively formal learning environment. The children, aged 8–9 years, included a high number for whom English was an additional language, including newly immigrant. The impact evaluation required the use of a multimethod research approach for capturing data (using observation, interviews, data logging, video recording, a specially designed survey instrument, and action research). Data embraced children's music making, language use, technology use, and sense of social inclusion. The teacher and the pupils acted as "co-researchers," that is, helped to develop the application, to provide insights into their experiences of the JamMo from the perspective of users.

Throughout the yearlong classroom fieldwork, and despite the many technical challenges of working with beta-grade software, the participating pupils remained overwhelming enthusiastic about JamMo and its potential musical opportunities. Pupils clearly looked forward to their participation, whether in whole class, small group, and pair activities. The pupils were aware of JamMo's ongoing developmental status and of their important roles as both "beta-testers." Perhaps, as a result of this, they tolerated missing functionality and frequent software crashes and they made many helpful design and interface suggestions drawing from often extensive preexisting music and IT experience.

According to pre-post JamMo questionnaire data and analyses, there is statistical evidence that participants (including migrant participants and children with special educational needs) reported themselves to be more socially included after several weeks' JamMo sessions. Although the cause and effect remain less clear, nevertheless, classroom observation by the adults involved (teacher, teaching assistant, researchers) suggested that children were particularly collegiate in their approach to JamMo. In particular, many participants considered as less socially included by the teacher and the teaching assistant enthusiastically engaged in JamMo activity. Furthermore, many migrant children and children with special educational needs engaged in the activity. They reported (and were observed) to enjoy JamMo a great deal.

In addition to longitudinal studies, smaller focused evaluations were carried out:

a. how technologies such as JamMo might be used to promote social inclusion in a general sense
b. development of research methodologies that would allow children to rate perceptions of their own social inclusion
c. focus on the engagement and enjoyment of young immigrant children as well as on the ADHD target group, pursuing a particular focus on (i) the self-regulation and coherence of behavior in musical contexts; and (ii) the proactive and interactive processes that participants employed when using JamMo, along with the resulting musical products
d. how singing game functionality of the JamMo for 3- to 6-year-old children worked. The research took place within a nursery school environment to investigate its impact on musical behavior, social inclusion, and digital literacy.

In addition, JamMo was also evaluated on two separate occasions by a group of secondary users. These were drawn from professional community music circles, practicing in contexts such as prisons and young offender institutions, pupil referral units, clinical settings, special schools, units for adults with learning disabilities, arts outreach workshops, and mainstream schools. These participants were very positive and they particularly liked the children-designed interface and the facility of the software to enable musical game playing in network mode, as well as for the potential for remote control provided by the teacher server in a classroom context.

1.6 Lessons

Designing mobile applications for children through experimental multidisciplinary research (such as in UMSIC) provided a lot of valuable lessons. As the requirements were collected in distributed manner and true requirements management guideline was not adopted, the designing team appeared to have some requirements gaps (from technical viewpoint, see Figure 1.8). The objectives were ambiguous and were not really fully linked from objective to the product (i.e., the partners spoke about same issues, but the language just was not the same). It should be noted that management of such a research with so many different partners of different scientific fields is very challenging (Lesson C). The atmosphere was enthusiastic and cooperation was working well, and this is an important issue. However, due to differences in sciences and cultures of communication, maybe too much was expected (Lesson A) and all possibilities were not seen (Lesson B). In some areas, constrains and

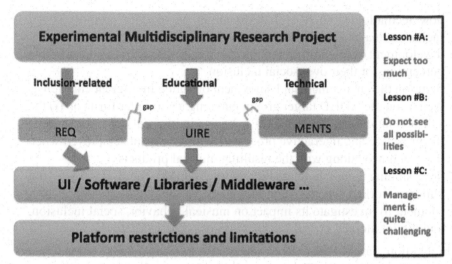

Figure 1.8   Lessons learned (technical view)

premade decisions too much guided the overall process toward the product. That might have left some gaps unfilled.

1.6.1 Lessons for Educator

Mobile technology and mobile software, particularly designed for children and music, are not common. The portability of the mobile device can be seen as a socially inclusive feature of the learning environment; specifically, children can pass the smart phone around and "gather round" it. Research fieldwork has suggested that it is inherently easy to share such a mobile device. Its portability means that it can be incorporated into standard classroom "show and tell"-style plenaries, that is, children can sit in a circle with their smart phones and listen and look at each other. Again, research fieldwork has suggested that in contrast to individual work on desktop computers, the size of the device encourages users to get closer to each other and that their body language affords more potential for interaction and "face-to-face" contact. Children participating in JamMo evaluation fieldwork activities were observed to like working in pairs or in small groups. The children were motivated to use smart phones in music making and it seemed natural for them to use the devices in such a way, promoting collaborative approaches to activities such as composing. Another significant inclusive feature of the software has been its accessibility: The fieldwork participants reported that the games were easy to use.

There remain several incomplete elements of the software that await future testing. It is anticipated that at least some of these elements include the potential for fostering social inclusion. For instance, the 7–12 game functionality has been subjected to only minimal testing; these games support learning as a basic premise of social inclusion. The networking functions of the software—facilitating musical communication with peers whenever and wherever in free time—may also support social inclusion. Additionally, the teacher software is planned to make it possible to group children according to their social needs, and to monitor and support individual children's learning processes. The possibility for wireless communication in sharing workshops may also make musical collaboration easier for children.

1.6.1.1 What worked well with target children and teachers?

Children participating in fieldwork research activities reported liking the idea of using phones for learning, even if reality offered a less satisfying experience (due to technical challenges and limited functionality). In particular, the children liked the graphics and the interface. Most said that they saw clear links between the icons, interface, and musical materials (i.e., stylistic congruence). Despite possessing a different kind of interface, JamMo's "look and feel" has been observed to be close enough to other music software for children to understand it with very little prompting or explanation. The basic premise of constructing music using preexisting musical loops was clearly attractive to children, and research has suggested that they enjoyed the musical materials of JamMo. In several cases, there was evidence of dancing and obvious physical responses to the music. The children also appeared to like the 3–6- and 7–12-composition game concept (understood to be a simple track-based sequencer). Moreover, 7–12 workshop activities, in which groups of four pupils used the sequencer, were also deemed to have been successful. While aimed at a younger age group, researchers have found that children of up to 11 years have also been able to engage creatively with the 3–6 version of the software. However, those in the 10–11 age group often reported finding the product "too easy." Nevertheless, the children generally felt that they had accomplished composition tasks well with both games. Terms used by the pupils participating in research fieldwork to describe their compositions included "fine," "nice," "beautiful," "OK," "rock," yet also "strange."

Teachers whose pupils were participating in research fieldwork have reported that both the younger and older composition games appeared to provide suitable learning materials children up to 11 years. In particular, teachers reported that the software appears to be broadly appropriate for

UMSIC's two target groups, recently immigrant children and children with ADHD. According to many teachers, children demonstrated high levels of motivation to use JamMo. Those in Finland also regarded JamMo as having had a positive effect on the social interaction and collaboration between children with ADHD and their peers. Teachers also regarded JamMo as a useful teaching tool, and the software's provision of different levels in the pedagogical design was reported as a positive feature. Teachers have also commented favorably on JamMo's clear interface and extensive library of musical materials. The teachers also showed enthusiasm to use JamMo or similar software in the future.

1.6.1.2 Music educational experiences

Again following fieldwork, researchers have been reminded that the musical materials provided on games such as JamMo (specifically loops and backing tracks) have to be of good quality to keep children engaged and excited, since many young people's expectations are very high these days. This had been anticipated during the design phase of JamMo and, while recording live musicians performances and editing musical materials formed a lengthy process, it was certainly worth doing.

Even though the JamMo software was limited by technology at times, it was useful as a means of promoting high-quality discourse and reflection on music. Researchers reported having had a number of engaging conversations in the classroom with a high standard of musical vocabulary and level of analysis. Children respond very well to being involved in the educational process as "coresearchers" and beta-testers. Researchers were clear that the software remained unfinished and that children's feedback was valued. As a result, they tended to overlook crashes and bugs and offered many good ideas and suggestions. Fundamentally, they were pleased to be involved in exploratory activities.

Our experience has suggested that there are some basic educational issues that should be considered in the design mobile music applications: (1) Schools may differ in their learning cultures, and these differences are reflected in mobile learning. Some researchers have found that children may need to be ready to collaborate in general within their learning before being ready to collaborate specifically with mobile devices; (2) other researchers have articulated a need for clear task instructions to precede action, and for a clear rationale to exist as to whether children work with musical instruments or mobile devices; 3) many children appear to respond enthusiastically to compositional tasks, particularly when they are structured and formed in suitable portions.

It seems that children are used to self-assessment and monitoring in their learning and that this is naturally extended to their experiences in mobile learning. Moreover, our research fieldwork has suggested that teachers are much more ready to use new learning environments than some years ago. As a result, we would suggest that there has been a positive change in attitudes toward new technology in music education. Tips for music educators for developing mobile applications are collected in Table 1.2.

1.6.2 Lessons for Mobile Application Developer

As the development was carried out in distributed teams, the management became an important issue. A head developer was chosen from the beginning

Table 1.2 Tips for music educators for developing mobile applications

Tips for Educators
• See the software in use with children—many problems can be identified/solved very quickly this way
• Keep a good communication channel between the researchers in the classroom and the development team, and all other stakeholders.   NOTE! Also important for developer
• Have a multidisciplinary team whose expertise can cover the whole music–education–technology–software development-research cycle.   NOTE! Also important for developer
• Study interaction methodology in a multidisciplinary context
• Specify in detail, what children are expected to learn about music and with the software, and how each of the games correspond to these aims
• Construct games and interactive settings that do not need any verbal instructions that are implemented within the software
• Do not try to make your tool do too much. Keep it simple and make sure the technology can support your aspirations
• Touch screen delay is too long for making music in real-time
• Start with classroom realities and work out how the technology can assist/support—do not start from the technology and then "impose" this on the classroom context
• Involve children in the design and evaluation process
• Remain flexible—be aware that your dreams and aspirations will have to be tempered by the realities of the technology, time available, and classroom/curriculum requirements
• Technical preparation for music lessons takes time
• Educating the teacher to use the software in a pedagogical situation is essential, and the teachers' voice is important to be listened to
• Teachers should get familiar with the software before the lessons, and prepare the lessons well

that eased some of distributed flow of operation. However, real centralized management for the integration was not established from the beginning and the teams were not fully aware what module should be ready by what time and what information would be required from others at what time. As the project was trying to reach too much, in sense of, for example, device capabilities, a lot of ideas and concepts were discharged. It would have been beneficial to be able to test the equipment performance prior implementation (lesson D). Another major lesson (E) was related to the programming itself; some of the decisions, for example, selecting a certain library should have been studied more in depth so that its effects on the development would have been known. Closely related to that is another lesson that all these decisions should have been made already in the beginning of development. As in most projects, the communication between distributed teams should be "agile" and exhaustive enough (lessons F and G). All development activities should also be transparent, so that the motives and reasons for selections are clearly shown, lesson (G). It should be noted that the focus (on development) is on continuous process improvement, rather than a waterfall model in which progress is seen as sequentially flowing downward process (like a waterfall) through several phases. Table 1.3 shows the summary of lessons learned for mobile software development.

1.6.2.1 Challenges related to software and hardware

Overall, the research fieldwork conducted to date has suggested that the most important requirement related to ICT is having powerful enough computing resources to facilitate the chosen learning objectives. At present, the selected mobile device is not powerful enough to perform all of the anticipated pedagogical designs.

Project fieldwork has reminded researchers of the need to test educational software for long enough and as many times as is required for the product to be become mature. This is a particular issue for software intended for use with children with learning difficulties. Software stability was an issue throughout the much of the fieldwork conducted to date. Some children criticized the smart phone version of JamMo for being too slow, crashing suddenly, getting stuck, and not saving their products. The laptop version was much found to be much more stable; however, there were still crashes and this did disrupt both education and research. The random musical subtheme selection in the JamMo 3–6 composition game (i.e., when returning to a "city" composition theme, children would not necessarily be presented with the same loops as the last time they used the screen) made discussion work hard. This was because teachers and pupils could not compare the themes back and forth. A further key

Table 1.3 Lessons learned for mobile software development

Lesson	Comment	Development Tip
Lesson D: Platform and equipment requirements, constrains, etc. need to be tested prior to implementation	Do not overestimate device capabilities! Mobile devices are still limited in their resources	Test the actual performance before starting the implementation of real product
Lesson E: Software and programming-related decisions should be based on well-known principles and best-of-practices (aligning with lesson D)	Selection of programming languages used, libraries used, etc.	Test the libraries on the selected device before selection
Lesson F: The project management and the management process of the requirements should be clear and explicit	On large-scale project change management becomes vital. The decisions need to be made in controlled way and information needs to be distributed to all parties involved in project—developing or using the software	One place where all information is available. Information needs to be shared—especially the changes.
Lesson G: The overall software development process and its activities should be transparent throughout the development life cycle	It is not enough to know what will be done, the reasons behind all decisions is also needed	Track the source of all requirements; mark who is responsible for implementing the solution for them and when they are implemented

requirement highlighted during fieldwork was the need to be able to save and retrieve products from one session to the next. Some children had difficulties in the singing game 3–6 since it was not always clear when they should start singing. Additionally, children singing unfamiliar songs required additional support in learning the lyrics. A further requirement identified during the course of the fieldwork was a function to change the pitch and tempo to suit voice type. Some children did not find important navigation icons during work with the 7–12 composing game. The teachers criticized the presence of concepts perceived to be unfamiliar for pupils such as "a sequencer" in JamMo. They also reported the smart phone touch screen to be too insensitive. Table 1.4 shows the notes gathered for mobile software development.

Table 1.4   Notes made for mobile software development

Notes for Application Developers
• Get feedback from field tests with children—many problems can be identified/solved very quickly this way
• Keep a good communication channel between the researchers in the classroom and the development team—and all other stakeholders. NOTE! Also important for educator
• Have a multidisciplinary team whose expertise can cover the whole music–education–technology–software development-research cycle. NOTE! Also important for educator
• Apply professional software engineering know-how on how to manage such a process
• The software developers cannot always predict how long programming will take—make sure other stakeholders know that implementation schedule may change
• Clear graphics is important in a small screen
• The device capabilities limit what can be implemented. In our case, the music making with virtual instruments was too slow to be usable
• Technical preparation of field tests takes time

1.7 Conclusions

Designing for children, in particularly, with mobile technologies, is in very early stages, and well-established good practices are not yet available. In this article, lessons and impact of children-centric research project were introduced and discussed. It is important to note that such projects are highly multidisciplinary and the management is very challenging. However, if all parties are enthusiastic, marvelous results, as in UMSIC was the case, can be obtained. On the other hand, with such a setting for a project often all either expect too much or do not see all possibilities. In the area of technology, off-the-shelf technology and the already-established tools, libraries, etc. should be used to maximize the performance in the sense of reality. Main lessons for educator can be largely identified to issues such as (i) what worked with children and (ii) educational experiences. Impact evaluations of the projects' result, JamMo, indicate that there was evidence the children felt more socially included. However, further studies need to be carried out to clarify the causes and the effects of the JamMo. The project has greatly contributed to children-centric design both in general and in developing mobile application for children.

References

[1] Brederode, B., P. Markopoulos, et al. *pOwerball: The Design of a Novel Mixed-Reality Game for Children with Mixed Abilities*. IDC 2005, Boulder, Colorado, ACM Press, 2005.

[2] Druin, A. *Mobile Technology for Children*. New York, Morgan Kauffmann, 2009.

[3] El Amam, K., and Madhavji, N. Measuring the Success of Requirements Engineering Processes. In Proceedings of 2nd IEEE Requirements Engineering Symposium, York, U.K, IEEE. 1995.

[4] Gilutz, S., and J. Nielsen. *Usability of Websites for Children: 70 Design Guidelines*, 1–127. Freemont, CA: Nielsen Norman Group, 2002.

[5] Hart, R. *My Mobile: UK Children and their Mobile Phones*. Intuitive Media Research Services, 2007.

[6] Hood, C., et al. *Requirements Management*, Berlin: Springer, 2008.

[7] Inkpen, K. "Drag-and-Drop Versus Point-and-Click Mouse Interaction Styles for Children." *ACM Transactions on Computer-Human Interaction* 8, no. 1 (2001): 1–33.

[8] Kafai, Y., C.C. Ching, et al. "Children as Designers of Educational Multimedia Software." *Computers Education* 29, no. 2/3 (1997): 117–126.

[9] Liu, L., et al. *Why Requirements Engineering Fails: A Survey Report from China*. In Proceedings of 18th IEEE Requirements Engineering Symposium, Sydney, Australia, IEEE, 2010.

[10] Read, J.C. "Designing Mobile Phones for Children—Is there a difference?" *IJMHCI* 1, no. 3 (2009): 61–74.

[11] Read, J.C., P. Gregory, et al. *An Investigation of Participatory Design with Children—Informant, Balanced and Facilitated Design*. Interaction Design and Children, Eindhoven, Shaker Publishing, 2002.

2

Designing Mobile Applications
for the Construction Industry

T. Kallonen, A. Knutas, J. Ikonen and J. Porras

Innovation and Software Department, Lappeenranta
University of Technology, Lappeenranta, Finland

2.1 Introduction

Typical construction projects are complex, multi-partner projects creating unique products in varying locations. During a typical project, there are a lot of changes. Some are technical and some related to schedules. In any case, project partners need to be informed of the changes and they need to be able to react in a suitable way. Communication between different partners is extremely important to ensure successful project outcome and to improve the efficiency of all individual companies by reducing mistakes and costs.

The dynamic and mobile nature of construction projects makes them a viable candidate for the use of mobile applications. Mobile applications can be used on work site firstly to *collect* information, which might be of interest to other partners, and secondly to *share* the information to the work site/worker to whom it is related. As the application is aware of the processes, it can also give *guidance* to the worker related to the tasks.

Our research with the construction industry has focused on processes related to precast concrete elements (see Figure 2.1) in the element factory and construction site. In our context, the work in a project begins with construction planning. The plans for the building and individual elements are made and given to different parties in the project. With the plans, a schedule for the project is created. The schedule tells the element factory when the elements are needed on work site and therefore gives the deadlines for the manufacturing of individual elements.

Figure 2.1   Precast concrete wall elements on the construction site

Life cycle of a precast concrete element starts from design. The design tells what kind the element is (a piece of wall, floor, ceiling, etc.) and where it will be installed. Next the element production starts at the element factory. After it is finished, it will go through quality control and then either to storage at the factory or it will be transported to the construction site (Figure 2.1). There it can go to storage, or more often it will be installed directly from the truck. These processes are shown in Figure 2.2.

When an element production has finished, the element's main dimensions are checked as part of *quality control* to ensure that the element is manufactured according to specifications. A worker checks the main dimensions of an element using a measuring tape or laser dimension meter and fills the measuring records and signs them. The records are traditionally filed in a paper form and only used after this in error cases. After the quality check, the elements are transferred to storage or sent directly to work site for installation. The storage can be either inside a warehouse or outdoors. Typically, the storage and the storage locations of elements are managed by a single employee in charge of it and the information of the storage locations might not be available for other people. Usually this is not a problem, but finding the correct elements might be difficult if the person in charge is not available. Traditionally, the information about element *production status* does not move well between the element factory and the work site. Usually this is not even needed, as this information is only important in exception circumstances, if for example, the element is not ready according to schedule.

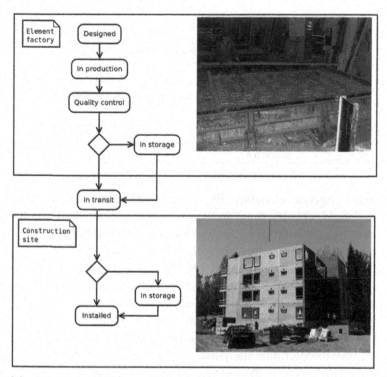

Figure 2.2 The life cycle of precast concrete element from the factory to the construction site

The construction industry has some problems that could be alleviated with the use of mobile technologies. One problem is the identification of precast elements. The precast concrete elements are traditionally *identified* by the identification code presented in the paper or plastic tag attached to a visible place on the surface of the element. Often within a building, there are several identical elements. All the identical elements share the same code so these codes are not unique within a project. Therefore, the code by itself cannot be used for tracking an individual element. The code also gets removed from the element during installation, so after the installation the element cannot be identified. Another problem is the fact that much of the information emerging during the construction process is only saved on pieces of paper or is not saved at all. Basically, these pieces of information are lost either immediately or at least when the project ends. These problems are summarized in Table 2.1.

To tackle the identification problem, the work presented here started on the *feasibility study of Radio Frequency Identification (RFID) technology* to

Table 2.1   The problems in the construction industry

Problem	Comment
Elements are not uniquely identified	Elements are only identified by type—identical elements have the same code
Records are in analog form	Much of the information gathered during the project, like quality assurance, is only on paper and is difficult to use especially after a long time
Status information does not move well between companies	The companies do not know the exact status of the production. Information is only transferred in case of problems

identify precast concrete elements. By using RFID tags embedded to the elements, we could identify them individually throughout their life cycle starting from the element factory all the way to the finished building.

The study continued toward *collecting and transferring information* related to the elements between the *mobile field workers* and other people involved in the same project. With our system, the workers get the element-related information *to work site* and the new information *from work site* gets sent to other project participants. The information is available where and when it is needed and it gets gathered and shared immediately as the work advances. The problem was to determine the *feasibility of mobile applications in the construction industry*. And if the use of these technologies is feasible, *how should mobile applications and information systems be implemented so that they offer more efficient and accurate information exchange than manual methods?* How can we make them usable to all users in a way that the use of these devices and applications will not distract manual work? We present the features of the construction industry that affect the design and implementation of mobile applications and give instructions and recommendations for application development on that field.

The results presented here were achieved through interviews of the employees in the construction industry (both precast element factory and construction companies in charge of the actual building process) and practical tests in the construction industry.

2.2 Information Technology and Construction Industry

Even though the work within the construction industry is highly mobile and involves a lot of collaboration, the construction industry has been fairly slow in adapting IT solutions as part of their processes. The communication still happens through personal communication, not by using common information systems and services. Due to the nature of construction, the use of IT solution could provide several advantages, but as an environment it is extremely challenging (May et al., 2005). The goal of the solutions should not be just to replace papers with mobile devices, but think of the advantages in a broader view. Some of the advantages of using mobile solutions on work site include reducing input errors, reducing paperwork, automatic generation of reports, and faster distribution of data (Cox et al., 2002). Despite the benefits, the use of mobile applications is still not widespread, but mostly in prototype and test phase.

Some work has been done using mobile applications within the construction industry. One such system is MOBIKO (Kirisci et al., 2004), which consists of mobile end user devices, a construction site server and a central server for gathering the work acceptance data from the construction site and sharing it to parties involved. The MOBIKO system proved to be successful in test cases but there were some crucial usability issues with the mobile devices such as ease of use, flexibility, and hands-free interaction.

Using mobile services to help the knowledge processes has also been studied within the construction industry (Skattor, 2007). Here the information was gathered from work site using mobile devices, analyzed and used to improve the overall performance of the processes of the construction company. The mobile services seem to be beneficial in comparison to gathering information in paper form and the end users seem to take more positive attitude to the system after they see the benefits of using it.

It seems that there is a lot of potential in the construction industry to benefit from using information technology and mobile solutions in its operations. The communication between workplaces and project partners has a lot of room for improvement and these improvements could benefit the industry as a whole. There seems to be a need for demonstrating that IT and mobile solutions can be usable in real cases to convince the decision makers of their applicability and thereby easing their way into wider use within the industry.

2.3 The Work Environment and Users

The end users of the mobile applications will be the workers at the construction element factory and at the construction site. These workers in general are not used to use computers or mobile devices in their work. They do use the mobile phones for ordinary voice communication and are familiar with the basic operation of the devices, but in general have not used mobile applications at their work. This brings some challenges to the design and implementation of the application. The application must be easy enough to be used at work site and its use must not distract the manual part of the work.

The workers in the construction industry from our point of view fall into two categories: the on-site workers using mobile applications and the office workers using web user interface. The office workers in general use computers in their daily work and their working environment is designed for it. The field workers on the other hand work in much more challenging environment with changing weather conditions and more dangerous work environment. Usually they do not have workstations available and use of laptops is not feasible, as they would not survive the harsh environment and are difficult to move around the workplace. The field workers in general are not so familiar with computers but they do use their mobile phones during work for communication. The mobile phone can also survive the weather conditions and other dangers in the work environment much better than bigger computing devices. Even though the mobile phone can be used in the construction fieldwork, it still has some difficulties. The use of keyboard or touch screen is difficult as the workers usually wear protective gloves. These need to be removed every time they need to use the keyboard; therefore, the use of keyboard should be minimized. These challenges are summarized in Table 2.2.

2.3.1 Advantages of Using Mobile Applications

The use of mobile applications in the construction industry has a few advantages over the old, manual way of working. First of all, when the data are collected using mobile applications and wireless devices, we can partly

Table 2.2   The challenges with using mobile solutions in the construction industry

Challenge	Comment	Tip
Weather, temperature, rain, snow, etc.	The weather produces challenges to both devices and people working in the construction industry	Test devices properly before taking them to work site. Make sure they work technically and can be used when wearing protective clothing
Noise and darkness	People use noisy devices on construction sites. The lighting conditions may vary	Do not rely only on sound to inform the users. Use other means as well
Change resistance	The users may not be familiar with mobile devices and unwilling to use them	Make sure the workers know the benefits and are given proper education on the mobile devices, applications, and the whole system

eliminate the typing errors by getting the data directly from another device. In the case of identification, the elements can be identified using RFID tags. Now the user does not need to type or otherwise input anything to the application. Measuring the element dimensions works in the same way. The dimensions can be received from the laser-measuring device in digital form and the user does not need to type anything.

Besides eliminating typing errors, the use of wireless data collecting devices has another advantage. We can be sure that *the work that is marked to be done is actually done*. In the case of dimension checks, it was previously possible just to mark that the dimensions were correct, even though no measurement was done. Now if we use the application on mobile phone and the wireless laser-measuring device, all measurements are collected from measuring devices and saved. There is now no easy way to skip the actual measurements.

Another advantage is *saving the data in digital form*. Previously as most of the data, like the quality checks, were saved in paper form, it was difficult to use the data afterward. Now when all data are saved on the server in digital form, it can be searched easily and can be used to analyze the work in a larger scale.

The mobile applications have also one major advantage when it comes to *communication* between different companies working in the same project. The status of one project party is easily transferred to another. For example, as the production statuses of single elements are saved to the database in real time, the construction site can easily check whether the element factory is in schedule. Or vice versa, the element factory can check the construction site status in order to see the installation and storage statuses of elements. This way they can prepare for changes in schedule faster and without manual communication. The advantages presented here are summarized in Table 2.3.

Table 2.3 The advantages of using mobile applications

Advantage	Comment
Reduce typing errors	When information is collected automatically, the user errors can be reduced
Mobile reliable information	When the quality checks are done with wireless devices, we can trust the results more than just a manual "all ok" marking
Digital data storage	When the data are digital from the start, it is easier to start and use after a long time
Automatic communication	A lot of the collected information can be sent/made available to other people in the project automatically reducing the need to spend time with communication

2.4 Mobile Application for Data Collection

One of the major software parts created in the project is the mobile application for Symbian series 60 mobile phones. With this application, it is possible to do the following:

- Identify elements using RFID.
- View the element information.
- Update element status (work started, finished, in transit, installed, etc.).
- Measure the element using Bluetooth-enabled laser-measuring device.
- Set and view element location.
- Create fault reports related to elements.

The goal in our projects regarding the end users was to provide easy-to-use applications and user interfaces which would guide the users through their tasks with simple user interfaces that give clear instructions on what the user can and should do during the task. For the user interface design, we followed task-centered user interface design (Lewis and Rieman, 1994), which is an iterative process that starts with the analysis of the users and the tasks they are doing. The user interface is then designed on the basis of these tasks and it should only have features that are needed in the tasks. The resulting user interface will be simple with only those features visible that can be used at that stage of the task. The user interfaces for different tasks are different and they follow the task structures of real world. The user interface gives the worker instructions on what to do next, so the task will be done in a correct way.

As the tasks are focused on the precast concrete elements, the tasks usually start with identifying the element using RFID like presented on the left part of Figure 2.3. The identification with RFID is a simple task. The user reads the ID from RFID tag with external RFID reader and the reader sends the ID to the mobile phone. The mobile phone receives the ID and can start the next task with the concrete element recognized. The center part of Figure 2.3 shows the details of the identified element, and the right part shows the menu for connecting devices or accessing different tasks.

The task to be started can differ between different users. For example, if the user does quality control, then the user interface for measuring the element dimensions should be started. In the presented Figure 2.4, there is a chart that shows the task structure: After the mandatory identification of the element, the program can show the user one of the several available interfaces depending on the task at hand.

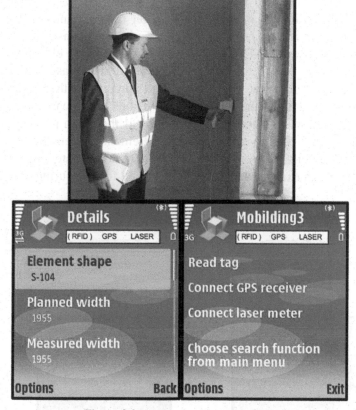

Figure 2.3 Examples of user interface

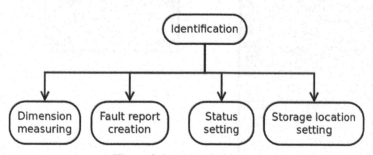

Figure 2.4 The task structure

2.4.1 Quality Control

In quality control, the precast concrete element is checked for visible defects and its main dimensions are measured to make sure they are within tolerances.

Our application can be used to measure the dimensions with the help of external laser dimension measuring–device. After the element is identified using RFID, the application retrieves its designed dimensions and allowed tolerances from the server and asks the worker to measure the element. Only the main dimensions (width, height, and thickness) are measured and the application always tells what to do next. The application then compares the results with the design (the designed dimensions are downloaded from the server) and displays a warning to the worker if there is a problem. The results are also sent to the server. Measuring an element happens step by step and the application tells the user at each step what to do. The steps are shown in Figure 2.5, and an example of user interface used for measuring is shown in Figure 2.6.

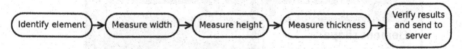

Figure 2.5   Steps for measuring an element

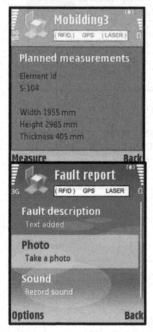

Figure 2.6   User interfaces for showing planned measurements, measuring one dimension and creating a fault report

2.4.2 Fault Reports

Another case of quality control is sending an error report. In any stage of the construction process if a worker notices a problem, he can create an error report with our mobile application (see the right part of Figure 2.6). Like other tasks, creating an error report starts with identifying the element using RFID. After the element is identified, an error report can be created. An error report can be a text description of the situation, a picture of the problem, and/or a recorded spoken description of the problem situation. After all the wanted data for the report are collected on the device, it will be sent to the server which after getting the report sends an email notification to persons who need to know about the problems of that particular project.

2.4.3 Production Status

Production status tells the status of the element from design to installation like presented earlier in Figure 2.2. Sharing the production status between project participants can help them follow the schedule and is especially important in exception cases. By sharing the statuses, other parties can more easily react to changes. With mobile application, the status of an element can be updated by reading the identification code from the RFID tag and selecting a new status from the mobile application (Figure 2.7). The status update can also

Figure 2.7   User interface for changing element production status

happen automatically along with other tasks. For example, setting a storage location changes the status to *in storage*. The system has the following statuses available starting from design to ready building:

- Designed
- In production
- In storage
- In transit
- Installed.

2.4.4 Storage Information

After the quality control, the elements are typically stored in the element factory before transporting them to the construction site. The storage can be inside or outside, and the storage locations are usually marked with a code like storage A1.

The mobile application can be used to save or view the storage location of the element. The storage location can be either a code of the storage area or coordinates from GPS device in the case of outdoor storage. In the case of storage location code, the RFID tag of the element is read and the storage location is selected from a list. At the same time, the status of the element is changed to *in storage*. In the case of GPS coordinates, after reading the RFID tag, the user only needs to select that the element is stored here, then the application will get the coordinates either from the phone's internal GPS receiver or from the external GPS receiver using Bluetooth. The storage location can be searched with the element code because here we cannot use the RFID as we cannot read the tag. The system displays the locations of all the elements sharing the same code on the map like presented in Figure 2.8.

2.5 Lessons Learned

The construction industry is a challenging environment for mobile devices and applications. First of all, the environment in the workplace is challenging for the devices with changing temperature and weather conditions. The devices should be designed in a way that they can survive this environment with changing temperature and moisture as well as the danger of being dropped on concrete and other physical harm. In practice, we encountered problems when reading RFID tags in winter conditions. It became more difficult to read the tag successfully when the temperature was around -20°C. Besides the problems

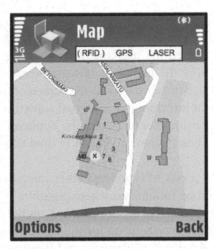

Figure 2.8 Map showing the location of the desired element

with the cold weather, the RFID technology proved to be feasible, but the devices would need some changes to be usable in real use cases.

As the workers move often from one place to another, the devices should be small enough to be easily carried. The mobile phone by itself is small enough but in our tests we used up to three external devices as well: RFID reader, GPS receiver, and laser dimension–measuring device. This is way too much to be practical. The GPS receiver is now commonly integrated with mobile phones but getting rid of other devices is more difficult. Some mobile phones are equipped with RFID or near-field communication (NFC) reader, but their read ranges are smaller than with external readers. In practice, they require the tag to be touched with the phone, so they cannot be used in our use cases where the tags are embedded inside the concrete. The need for external devices is a challenge also from application development point of view, as the usability of the whole system becomes a challenge. One thing to do is to minimize the need to use several devices at once. The worker should only use one device at a time so if the worker is using an external device, the mobile phone should not require any interaction. It should only collect the data received from the external device.

All the users of mobile applications in the construction industry are not familiar with using mobile devices or computers during their work, so the applications need to be designed with this in mind. First of all, the start-up of the application needs to be simple. The program should start with the correct view for the worker in question, and all the required connections with external devices should be started automatically. The application needs to be simple and informs the user on what the user needs to do next. To keep things simple, we started the user interface design by inspecting the existing processes. This way the use of mobile applications has a minimal effect on the work. To reduce the need to view the screen of the mobile phone, sound was used as a way of informing the user that something has happened. For example, during dimension measurement, the application plays a sound when one dimension has been measured and it is time to move to the next measurement.

User input has always a risk of typos and other mistakes. Therefore, the need for user input should be minimized where possible. If the data needed can be collected without the need to type something on the device, it should be collected that way. To reduce the need for user input, we used the RFID to automatically identify the concrete elements and Bluetooth-equipped laser dimension–measuring device to collect the element dimensions automatically. Other ways to minimize the user input (with keyboard) were the use of camera and microphone when collecting error reports. The user can take a picture of the error situation to remove the need for textual error description. In the same way, the user can describe the error situation by speaking using the microphone.

Table 2.4   Summary of the lessons learned

Lesson	Comment	Tip
Device reliability needs to be tested	Wireless devices may not be reliable in all weather or other environmental conditions	Test devices properly before taking them to the work site
Start-up needs to be easy	Starting the use should be easy. No setup needed	Set up the program so that it starts with the correct view for this employee. Set up all connection needed automatically
Limit the amount of typing needed	Typing creates typos. Typing requires effort	Collect as much data as possible with other means than typing. Use sensors, camera, microphone, etc.
Use sound or vibration in user interface	The use of sound or vibration reduces the need to see the screen while working	When the user is just informed that something has happened, sound or vibration can be used to reduce the need to see the screen

The use of mobile devices can be difficult because of the environmental conditions. The cold weather and harsh environment force the users to wear gloves, and the lighting conditions differ from direct sunlight to darkness. These conditions make both typing and viewing the screen difficult and therefore also favor automatic means for data collection and the use of sound and vibration in the user interface.

2.6 Conclusions

The construction industry is a challenging field for the use of mobile technologies and applications. The environment is challenging for the devices, as they need to survive the mobile work in harsh environment with changing conditions. The users in the field are typically not used to working with mobile devices. They do use mobile phones for communication, but not for anything else. Despite these challenges, the use of mobile applications does offer some advantages over the old, manual way of doing things. Projects in the construction industry typically have many companies working together to create a single product, so communication between these companies is important. Collecting information directly from the field in digital form makes it possible to share a lot of information to others automatically. The use of mobile applications has some advantages both inside a single company and especially when communicating with other companies in the same project. A single company can take the advantage of the digital data, as the reliability of the data is higher than in the case of paper files where typos are common. Secondly, analysis of large data is a lot easier when it is available in digital form. The use of mobile applications makes the communication between companies easier as the data collected can be automatically shared to others. The most important use of this is the sharing of production status—especially when automatically compared with the schedule. If something is not done within a schedule, information of this can be automatically sent to whomever it affects.

In our work with the use of mobile applications in the construction industry, we have shown that RFID and mobile applications *can* be used to collect and share information inside a company and within a project. There still are a number of challenges though. First of all, the devices, mainly mobile phones, are not designed for such an environment. They typically are not tough enough for the environment and not easy enough to be used with gloves on. Secondly, we needed several individual devices, the mobile phone, RFID reader, GPS receiver and a laser-measuring device. These would need to be combined into a single device to be practical, as we cannot expect a user to carry so many devices with him. The use of these external devices has huge benefits as they reduce the need for manual work while automating data collection, but carrying each of them individually makes the use cumbersome.

The mobile application we created proved to be adequate for the job as verified in the practical field tests of all the features. The overall idea of task-centered approach seemed logical for the users, and no major usability issues were discovered. The use of camera and sound recording was good for usability as it reduced the need to type on the keyboard and the use of sound and vibration in the user interface reduced the need to follow the screen when the user was focused on something else (like measuring an element with a laser-measuring device).

References

[1] Cox, S., J. Perdomo, and W. Thabet. "Construction Field Data Inspection Using Pocket PC Technology." In *International Council for Research and Innovation in Building and Construction*, CIB w78 Conference, 2002.

[2] Kirisci, P.T., H.-H. Hunecke, K.A. Hribernik, C. Dikici. "A wireless solution for mobile collaboration on construction sites." *International Workshop on Wireless Ad-Hoc Networks*, 166–171. 31 May–3 June, 2004.

[3] Lewis, C. and J. Rieman. *Task-centered user interface design: A practical introduction*, 1994. http://dcti.iscte.pt/cgm/web/TCUID_PI.pdf.

[4] May, A., V. Mitchell, B. Bowden, and T. Thorpe. "Opportunities and Challenges for Location Aware Computing in the Construction Industry." In *Proceedings of the 7th International Conference on Human Computer Interaction with Mobile Devices & Services* (MobileHCI '05), 255–258. New York, NY: ACM, 2005. http://doi.acm.org/10.1145/10857 77.1085825

[5] Skattor, B. "Design of Mobile Services Supporting Knowledge Processes on Building Sites." WCMeB 2007. *Eighth World Congress on the Management of eBusiness*, 11–13 July, 2007.

3

Innovation for the "Bottom of the Pyramid"—Mobile for Development Experiences of FarmerNet, Sri Lanka

H. Liyanage[1] and P. Edge[2]

[1]Sarvodaya-Fusion, Sri Jayawardenepura Kotte, Sri Lanka
[2]eNovation4D, Canterbury, Kent, UK

3.1 Introduction

Technology has always played an important role in the development. For instance, the impact of green revolution technologies to increase agricultural productivity in small land plots, and biotechnology applications for health, provides evidence for the power of technology to address issues of poverty. Developed country institutions, both public and private, have generally carried out technology development. In some cases, the results of this type of technology development have caused questioning of the "human face" of technology.

Information and Communication Technologies (ICTs) provide a new opportunity for viewing these issues. Although the origins of the basic technology, such as the computer, the Internet, and hardware, are founded in the developed West, the end-user application technologies are often a product of developing country inputs. This provides opportunities to tailor technology products to meet the needs of the poor.

In recent years, the mobile phone market in Sri Lanka, as in most other developing countries, has grown very rapidly. The potential for mobile technologies to influence poor communities in developing countries has excited the development sector whose primary focus is to help the poorest of the poor—the so-called bottom of the pyramid (Prahalad, 2004) which is generally defined as the largest, but poorest socioeconomic group, who live on less than $2.50 per day. A new sector has been defined—mobile for development (M4D).

Rapid progress in this sector is attributed to the fact that mobile technology has penetrated almost everywhere, and particularly to previously isolated rural areas, thus becoming accessible to poor rural communities. Sri Lanka's mobile infrastructure has penetrated these rural areas, where the highest percentage (82%) of poor people live. About 40% of these "bottom of the pyramid" people have access to phone communication. Thus, the mobile phone provides new hope and opportunity to support development objectives in which success has previously proved elusive.

Over the last few years, the world has seen continuous development of mobile phone-based applications, in line with development objectives in many developing countries. As communication technologies such as mobile phones become simpler to use and more standardized in design and functionality, the opportunities grow to develop services, which are tailored specifically to the needs and capabilities of people who may have very little previous experience of the technology. These applications now go beyond general voice transmission, ranging from disseminating market information, monitoring health care, transferring money, monitoring elections, and even to literacy and education delivery (Aker and Mbiti, 2010). Mobile phone services monitor measles outbreaks in Zambia and disseminate health education messages in Benin, Malawi, and Uganda.

While these developments provide evidence to prove the promise, they also unveil the challenges. Many of these challenges are associated with sustainability. The price of a mobile phone call can equal 40% of a household's daily income for a poor family in Niger. There are huge disparities in geographical signal coverage, and decisions about where and how to provide the coverage have been primarily decided by economic considerations (Aker and Mbiti, 2010). Such challenges have to be viewed in the context of the targeted end user and the environment that he or she is living in. At the "bottom of the pyramid" most of the people are poorly educated, if not illiterate, and their decision-making capacity is hampered by a plethora of problems as their lives are often exposed to shocks such as harvest failure due to pests or adverse weather, diseases, or political turmoil.

Because of the vulnerable nature of these communities, particular focus is brought to bear on sustainable development. As defined by the Brundtland Report (1987), "Sustainable development is development that meets the needs of the present without compromising the ability of future generations to meet their own needs," in which the concept of needs, in particular the essential needs of the world's poor, should be an overriding priority.

Within this broad view, the understanding of sustainable development can be seen from different perspectives, particularly economic, social, and environmental. This chapter elucidates the process of a technological innovation in a development-centric implementation, which targets the "bottom of the pyramid" while respecting the implications of sustainable development. The innovation is also shown to meet the needs of its users in the context of "appropriate technology," as initiated by Schumacher and promoted today by Practical Action (Practical Action 2010). FarmerNet, developed by Sarvodaya-Fusion (a Sri Lankan NGO), provides the core case study material for this chapter.

3.2 Organizational History and FarmerNet

Sarvodaya-Fusion is the information and communication technologies for development (ICT4D) arm of Sarvodaya, a 50-year-old national NGO in Sri Lanka. The programs of Sarvodaya are founded on a holistic development approach grounded in spiritual, moral, cultural, economic, and political dimensions of development (Premasiri, 1997).

Fusion's mission has been to design, develop, and adapt ICT technologies to address issues of the digital divide for poor rural communities, where illiteracy, poverty, and poor infrastructure present many challenges. Fusion has demonstrated its innovative capabilities repeatedly by designing development-centric ICT solutions targeting the rural poor as an integrated program within the broad range of ongoing community empowerment programs at Sarvodaya.

In the sectors of ICT4D and M4D, Fusion pioneered telecenters in 1997 and FarmerNet in 2009, both being the first initiatives of their kind in the country at the time of launch (see Table 3.1).

Table 3.1   Types of innovation by Fusion

Time Period	Innovation	Sector
1997–2000	Design, development, introduce telecenters as a development model	Telecenters
2004	Design and develop Subsidy Vouchers for telecenters	Community participation at telecenters
2005–2007	Virtual Villages project	WiFi technology application
2005–2008	Conceptualize and implement telecenter family network	Telecenter networking
2007	Design and development of ICT Book	Telecenter-based service
2007–2009	Design and development of ICT national exam	Telecenter-based service
2007–2010	Design and prototyping—FarmerNet (mobile + telecenters)	Mobile phone and telecenters

3.3 FarmerNet and the Impetus for Innovation

FarmerNet is a mobile phone-based application, which enables rural farmers to interact directly with traders, thus giving them control over the sale of their produce and the resulting income. It is a simple-to-use application, which can access all of Sri Lanka's mobile networks. The service was launched in August 2009, as a prototype, and was Fusion's first product development in the mobile application sector. What was the impetus for this mobile sector innovation? Has it followed the logical sequences of the innovation process as identified in current theory and practice? The thinking driving the development of FarmerNet is described in the sections below.

Bolton and Thompson (2000) explain "identification of a need" itself as a trigger for innovation. In the case of Fusion, it has identified a need and developed a solution—FarmerNet. The need is the absence of sufficient tools to improve the bargaining power of the rural farmer in Sri Lanka. Poor rural farmers in general do not have sufficient market access and as a result do not have an influence over the prices achieved for their produce. Less than 15% of households in Sri Lanka report having access to markets from their local communities (World Bank, 2006). These deficiencies in infrastructure pose barriers to the incremental empowerment process offered by micro-finance institutions (MFIs) (Charitonenko and de Silva, 2002).

The impetus for innovation, which led to the creation of FarmerNet, came from a complex of factors. As advocated in the innovation literature (Burns, 2007) and Drucker (1985) provides a framework for understanding and organizing such factors. In his book "Innovation and Entrepreneurship," Drucker lists seven sources of opportunity for entrepreneurs in search of creative innovation:

1. The unexpected
2. The incongruity
3. The inadequacy in underlying processes
4. The changes in industry or market structure
5. Demographic changes
6. Changes in perception, mood, and meaning
7. New knowledge.

For Fusion, the major focus since 1997 had been on the telecenter sector. Most of Fusion's work was in the operation of telecenter facilities, stimulating community participation, and facilitating telecenter sustainability. Nevertheless, as explained in Table 3.2 below, "the unexpected" poor community response, "the incongruity" of community readiness to invest in the telecenter systems, "the changes in the market structure" caused by rapid development and ubiquity of mobile technology, "changes in the perception" among the key stakeholders such as donors, and "the new knowledge" of the growing body of research studies have provided the impetus for FarmerNet innovation.

Table 3.2 illustrates and elaborates the comparable seven sources of innovation for Fusion, which turned its search for a solution to the problem (i.e., the need noted above) from telecenters to mobile applications.

Table 3.2   Sources of innovation leading to FarmerNet

Sources of Innovation	Comment
1. The unexpected	Recurring telecenter sustainability issues that had challenged the sustenance of the sector. Telecenters were not achieving the levels of usage within communities that had been hoped for. Further, the costs of infrastructure development and maintenance were high.
2. The incongruity	– Village communities were not ready to invest to build telecenters despite their appreciation of the technology. High expenditure and infrastructure requirements resulted in higher risks that discouraged small investments.
	– Micro-loans (SEEDS)* did not recognize the telecenter as an economically viable micro-enterprise model, because pilot feasibility experiments had failed to produce convincing results.
3. The inadequacy in underlying processes	Because levels of IT usage within the poor communities were low, it was difficult to develop a sufficient number of telecenter-based services to generate satisfactory impact in the context of broader development objectives.
4. The changes in industry or market structure	Unexpected and rapid development of mobile technology and ubiquity in and around the rural sector.
5. Demographic changes	– Over 90% of the rural population uses mobile phones.
	– In contrast, less than 5% of adults participated in telecenters.
6. Changes in perception	Donors and partners started recognizing the mobile phone, because of its' ubiquity and accessibility, as a better ICT application for development.
7. New knowledge	Growing body of case studies and research evidence on the effectiveness of M4D applications.

Sources: Observations, SWOT analysis records, strategy documents (*SEEDS is Sarvodaya Economic Enterprise Development Services)

3.4 Idea Generation to Prototype Development—the Innovation Process

How did Fusion transfer such impetus into a product concept? What was the product development process applied to FarmerNet? Does it follow the logical sequence observed more widely for innovation processes? Most innovation models are linear and unidirectional, beginning with idea formulation and following through to product launch (Bessant and Tidd, 2009). One model, the stage-gate process, provides an explicit process of six stages: idea development, preliminary investigation, building business case, prototype development, testing and validation, and full production and market launch (Cooper, 2000) (see Figure 3.1). An important feature of this model is the "stage gate"—at which point the deliverable (output) of the previous stage is assessed against a set of criteria and only if it qualifies will the process continue to the next stage (Cooper, 2000; Stamm, 2008).

FarmerNet development stages illustrated in Figure 3.2 follow a pattern that fits well with the stage-gate process from idea generation to prototype development. Further analysis describes the "stage gates" that have been applied to qualify each stage.

The three stages of idea scoping, concept development (or concept building) and prototype development, and the involvement in them of end-user perspectives and other stakeholder needs, are described in more detail below.

3.4.1 Stage I—Idea Scoping

Understanding the specific needs of the rural communities was the foundation of idea scoping. It was a gradual process of accumulating knowledge about the barriers the communities face in relation to wealth creation (educational, agricultural, social, infrastructural, medical, and so on). Sarvodaya's field workers are in constant contact with the target group (the rural community).

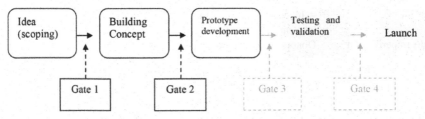

Figure 3.1   FarmerNet is currently at a stage between prototype development and testing and validation stage of the stage-gate innovation process (introduced by Cooper, 2000)

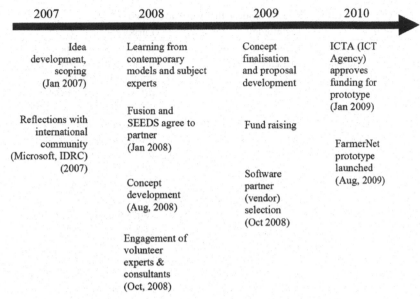

Figure 3.2 FarmerNet development stages and timeline
Sources: Reports, Project proposals, Meeting minutes

They live in the local environments, and interact with the target communities on a day-to-day basis, while implementing a diverse set of development programs. These authentic interactions mean that the field workers accumulate a rich tacit knowledge. Fusion carried out brainstorming sessions with these social workers as the core strategy for understanding the localized needs of the target community.

Through such brainstorming sessions, the Fusion team, the small group of internal staff members involved in the innovation process, initially developed five ideas for M4D developments for these user groups (Table 3.3). Though the ideas were quite diverse, they have two common characteristics: (a) the urge to combine mobile sector applications with telecenters, and (b) a purpose to serve the development needs identified for the bottom of the pyramid communities.

In order to stage-gate the idea that would eventually become FarmerNet, the following criteria (Table 3.4) were designed and applied. These criteria bring into focus factors in both the internal and external environments in which Fusion operates as a social enterprise. For instance, strategic fit and resource intensiveness are relevant to the mission and internal capabilities of Fusion, whereas technical feasibility, risks, and returns are most relevant to external factors.

Table 3.3   List of ideas that were developed

Idea	Technology	Purpose
Mobile phone application to improve pest and disease information dissemination	Mobile phones and telecenters	Dissemination of pest and disease information, combining with telecenter infrastructure for rural farmers
Application of mobile, PDA, and GIS technologies for snakebite prevention	Combining mobile phone, PDA, and GIS technology	Dissemination of health information for snakebite prevention to the rural sector
Application of "community PDA" for livelihood development	Combining mobile phones, PDA, and GIS technology	Provision of support tools (decision support, educational and income generation) to rural community
Convergence of Facebook and mobile phone for community empowerment	Mobile phone and social networking (Internet, Facebook)	Peer networks and village networks using Facebook as a back-end support system
Mobile phone application to facilitate rural farmer trading, using rural telecenters as information centers	Mobile phones and telecenters	Support rural farmer trading

Sources: Concept drafts, donor proposals, emails.

Table 3.4   Criteria applied to the idea that would become FarmerNet

Criteria	Comment
Strategic fit	Does the idea or concept fit into Fusion's mission, strategic objectives and capabilities of staff?
Technical feasibility	Is it technically feasible to develop within the developing country context in terms of acquiring services from available in-country software vendors?
Resource intensiveness	How feasible is it to manage the project within the limitations of Fusion's fund raising and fund managing capacity?
Risks	What is the chance of failure due to competitive forces from corporate telecoms and other potential competitors?
Returns	How effectively does it address the needs of the end-user communities? How feasible is it to develop the idea into a marketable product in a social enterprise context?

Sources: Personal notes, meeting minutes, reports.

The concepts shown above in Table 3.4 were not at this stage directly exposed to the intended end users (the farmers) because they were both somewhat intangible and conceptual and were difficult for these intended beneficiaries to grasp and evaluate as being of direct benefit to them.

In order to refine the idea and the stage-gating criteria, input from external expert communities was also gained at this stage. Experts included international donors (e.g. IDRC Canada, Microsoft-UP), corporate sector research organizations (Microsoft Research Group, Seattle), local think tanks (LIRNEasia, Sri Lanka), and subject experts (UK- and Canada-based experts on the Internet, mobile applications, and ICT4D development). These meetings were mostly consultative meetings, where Fusion team members engaged with these experts on an individual basis, either face-to-face or other means, to present the ideas and seek advice.

3.4.2 Stage II—Concept Development

With FarmerNet fitting well with the criteria applied, both internal and external, it was agreed the concept would be developed. It was decided that FarmerNet would be a mobile phone-based application, with simple, cross-platform functionality, which would connect farmers in real time to commodity prices and traders in the marketplace. By having direct access to market (instead of depending on middlemen in the trading chain), the farmer would become empowered by having direct control over the traded value of his/her crops. It was also seen to be important that the use of FarmerNet would require only basic skills, and to this end, the functionality would need to be built around just a few SMS text-based codes, not requiring the input of lengthy word sequences.

This concept development stage involved developing the idea into an executable concept by exploring similar existing product concepts, gathering specialized expert knowledge, and identifying potential partners who could contribute and add value. TradeNet, developed by BusyLab of Ghana (Kutsoati and Bartlett, 2008), was subject to detailed study, where its functional elements and the technologies employed were assessed. Expert consultations included meetings with advisors with in-depth experience of the M4D area. These meetings were mostly brainstorming meetings held in Sri Lanka and international locations, where graphical illustrations and modeling were employed to develop the ideas from meeting to meeting.

These key investigations included the following areas:

1. Technological concepts

   a. Technological models to harmonize mobile and web platforms.
   b. Use of open-source technologies versus proprietary software applications.
   c. Multimedia application versus simple (SMS) text messaging.
   d. Limitations of the mobile hardware (because the target community tends to adapt low cost, basic products, and mostly older models).
   e. Telecom operator issues such as geographical coverage, business models, Corporate Social Responsibility experience.

2. Target end-user responses

   a. Breaking the barriers of illiteracy and low technology exposure (i.e., the majority of farmers are poorly educated, if not illiterate, and the least exposed to ICTs).
   b. Breaking the traditional middleman barriers (most traditional small farmers continue their trading through middlemen).
   c. Individual participation versus group (society-based) participation (i.e. the majority of small farmers are members of farmer groups, village societies, etc.)
   d. Mobile owner versus mobile user participation (i.e. farmers may often use mobile phones, though they may not own one).

The outcome of this stage was the initial concept framework for FarmerNet with the key features of mobile SMS texting, software platform selection, key template elements, and spot trading logic. Figure 3.3 presents the detailed illustration of the concept.

In order to assess the validity of the stage-gate criteria used (as shown in Table 3.3), the concept was presented to two different groups: (a) target group (rural farmers), and (b) technical experts. Participatory methodologies were used to gain feedback from the rural farmer community. The concept was introduced to two rural famer groups (of 20 participants each) in two separate villages. Initially most farmers could not understand the concept, as they were only familiar with the conversation aspects of mobile phones and not data transfer and SMS texting. Dummy models and canvas drawings were used to introduce the concept to farmers, where role-play of farmer–trader interaction was also carried out. This approach helped the farmers to grasp the concept. Feedback from the experts was based on detailed face-to-face presentations, where the concept was presented in PowerPoint slides.

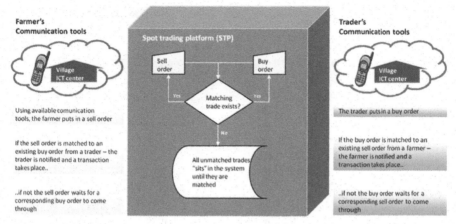

Figure 3.3   System overview of the FarmerNet spot trading logic
Source: FarmerNet concept presentation

3.4.3 Stage III—Prototype Development

Though it has been presented as a prototype, FarmerNet is a fully developed live software application that is functional through all the mobile networks in Sri Lanka. Nevertheless, it has been introduced to rural farmers on a controlled basis into selected communities in one district.

The FarmerNet prototype has the following key features:

- A simple web template that is easy to use and which targets rural farmers as well as traders (buyers, both small and large) to upload their selling and buying information either by using mobile SMS text messages or by online access via a telecenter (see Web site: www.farmer.lk) (Figure 3.4).
- Users register on the system as a buyer (trader) or a seller (farmer), and the system generates a user profile and password. After logging in to the system, the user may upload their information using an SMS message constructed through a sequence of eight codes.
- The "Spot trading logic" automatically searches for matching buyers and sellers based on six criteria including type, grade, and price. Matched partners are mutually informed by automatically generated SMS text messages, enabling them to carry out trading on their own. This creates a mobile-online marketplace where rural farmer and buyer can meet online or through mobile phone access.

The premise of the initiative is to create an efficient marketplace, using information technology to reduce transaction costs. The model leverages the

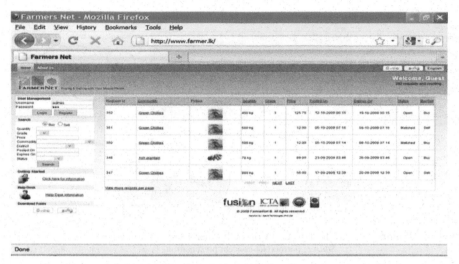

Figure 3.4 An image of FarmerNet web template (www.farmer.lk)

existing ICT infrastructure (mobile phones, telecenters, and Internet) to bypass the physical infrastructure barriers.

The FarmerNet prototype was developed by a local (private sector) company (Sabre Technologies Pvt. Ltd.) using open-source software. Sarvodaya Economic Enterprise Development Services (SEEDS Gte Ltd.) has collaborated with Fusion to develop the prototype, while ICT Agency of Sri Lanka has sponsored the prototype.

3.5 Ensuring Social Mission

Sustainable development has been defined in many ways. The Brundtland Report (1987) definition has already been referred to in this chapter: "Sustainable development is development that meets the needs of the present without compromising the ability of future generations to meet their own needs." There are many different views on what sustainable development should deliver, because it can be seen from social, economic, technological, and environmental perspectives. It is in the nature of Fusion's work, aims and mission that a balance should be found between these different perspectives, thus delivering long-term benefits to the poor.

In the context of technology application for sustainable development, E.F. Schumacher's book, "Small is Beautiful" (Schumacher, 1973), provided the early impetus that has fueled the development work carried out globally by

Practical Action, the international NGO. The numerous technologies promoted by Practical Action range from cooking stoves to hand pumps, and micro-hydros to solar energy. Practical Action defines six criteria to judge the appropriateness of a technology that:

- meets the needs of both women and men.
- enables people to generate income for themselves and their family.
- is affordable.
- has a limited impact on the environment.
- can be designed, improved, managed, and controlled by local people.
- uses local skills and materials as much as possible (Practical Action, 2010).

These indicators address the holistic view that the NGO sector in developing countries seeks to address. It corresponds with the context of poverty that resides in close affiliation to a particular cultural setting. Thus, the ability of FarmerNet to qualify, as measured against this comprehensive framework, can be recognized as a better way to assess its social appropriateness as a technology to support sustainable development.

The following section provides the body of evidence that was built using the above criteria to investigate the validity of FarmerNet as an appropriate technology to support a development purpose.

Gender neutrality: FarmerNet is built into mobile phone technology. Research evidence reports very low disparity between genders accessing phones in the "Bottom of the Pyramid" communities in Sri Lanka (Zainudeen et al., 2010). However, early field tests (of FarmerNet) indicate age bias (regardless of gender), where young people reportedly are more involved than the adult groups. This was mainly attributed to the familiarity with the feature of SMS texting. Adults tend to show reluctance and less comfort with texting, and often seek the help of children to do it on their behalf.

Income generation potential: The primary purpose of FarmerNet is to increase market access and improve the bargaining power of the rural poor farmer. This enables increased farmer income. Research evidence implies that the benefit of mobile phones can be transferred beyond mobile phone ownership, as multiple users can receive the services of a mobile phone rather than just the owner (Zainudeen et al., 2007). Thus, FarmerNet has the potential to serve the trading requirements of a group of people beyond the individual.

Affordability: The rural community can subscribe to FarmerNet without an additional payment, and the present model does not charge a fee for its services. Neither does it need any additional investment, as it does not require buying new hardware or software to use the product. The model was designed to work on the old generation mobile handsets as well as new ones. More importantly, FarmerNet was designed expecting to utilize rural telecenters as a supplementary back-end service to access the online platform. Thus, it promotes the application of existing mobile and telecenter platforms in the rural context, without requiring any new investment, which improves affordability.

Environmental impact: FarmerNet does not leave any specific environmental footprint; rather, it amplifies the efficient application of currently existing ICTs in the rural context. Nevertheless, it may have a downside in terms of its potential to increase ICTs in general, which accounts for 2.5% of global greenhouse gas emissions. However, Fusion has argued, this may be counterbalanced with the potential to cut down transportation needs and production losses that account for a significant proportion of energy waste in these rural locations.

Manageability and control: FarmerNet is not a capital-intensive product that requires proprietary software, extensive servers, or intensive human resources to manage. The technology is manageable within the developing country context, as local vendors have already developed it. Furthermore, the product operates well in all the available mobile telecom networks in the country and is not operator dependent. With these features, it is manageable within the NGO context (Fusion), which resides in the governance and control mechanism of Sarvodaya, the organization registered under a Government (Parliamentary) acts as a people-owned movement and governed by an executive body annually elected by village leaders.

Application of local skills: The functionality of FarmerNet is built around six SMS text-based codes and does not require the input of lengthy word sequences and hence does not require extensive literacy skills. Nevertheless, it requires the user's ability to input the codes using small keys in a small handheld device (the mobile phone). As noted above, early test results have indicated the difficulties experienced by adults (over 45 years of age) in SMS text messaging, remembering the codes, usernames, and passwords. Nevertheless, with reasonably simple assistance provided by the trained local staff, this can be overcome.

Does the appropriateness of the FarmerNet technology lead to real impact in the community? Does it make a difference? An impact analysis carried out by Fusion staff provides insight into this question (Fusion, 2010).

Table 3.5 Indicator-based assessment of FarmerNet's impact on an individual rural farmer as carried out by Fusion's social impact assessment study (Fusion, 2010)

Indicator	Narrative
Exposure	Exposure to FarmerNet and recognizing the ability to use mobile phones beyond voice communication purposes
Motivation	Interest to learn about FarmerNet and test it as a learning step
Skill development	Development of SMS texting skills and marketing skills using Farmer-Net with the assistance of trained staff
Self-learning	Self-learning to improve the skills to apply FarmerNet without assistance of trained staff
Exploration	Exploring the application of FarmerNet for marketing by repetitive posting and observing the response trends
Application	Application of the skills of mobile phones and FarmerNet for the benefit of the other community members

This impact analysis stipulates that the user goes through a journey from first-time exposure (to the FarmerNet application) to applying it for self-benefit and sometimes for the community at large.

The six stages of this journey that have been recognized are used to assess the process over a period of time, as indicated in Table 3.5. Each stage is shown in the Table as an indicator, with a narrative, which expands the context. These stages may not necessarily be linear and can be repetitive, but cumulatively develop the overall impact.

Impact occurs at three levels: (a) farmer education, (b) skill development, and (c) application for their own benefit and for the other members of the community. Education enables the farmer to recognize a new option to use the mobile phone for an agricultural trading purpose. The skill development collectively helps to enhance capabilities to use the technology. Application for self-benefits and community benefits enhances the influence beyond the self. Overall, this indicates the potential of FarmerNet to supplement the empowerment process that may be envisaged as the primary objective of the overall development engagement.

3.6 Conclusion

FarmerNet is a mobile phone-based application that works to carry out commodity trading between rural farmers and traders. The primary target group is the poor rural farmer, a "bottom of the pyramid" group in Sri Lanka. This group has previously experienced great difficulty in gaining direct access to markets to sell their produce. FarmerNet has provided an ICT tool to facilitate this direct access for the first time.

FarmerNet was developed, from the scoping of the idea to implementing a working prototype, in a methodical way which followed processes similar to those described in the literature on innovation theory and practice. And further, the innovation process involved stakeholders such as users (the farmer) and expert providers (software developers) in the concept development process. This study indicates the many challenges and alternatives that have to be addressed in engagements of this type.

The development of FarmerNet has taken place by addressing the needs of poor rural farmers and has been modeled to fit with both their social and economic environments as well as the technical requirements and limitations of their work and lives. The technological appropriateness of FarmerNet can be shown when mapped against the criteria for this used by Practical Action. FarmerNet was developed to address an existing social need of the target community while at the same time using innovation models more typically used by the private sector. This approach is starting to show signs of long-term success and sustainability in a situation where existing market forces alone have not so far been able to display such success.

The creation of FarmerNet shows that the detailed needs of both technical development and "fitness" for the individual, the community and the environment, can be met in a development engagement of this sort.

Bibliography

[1] Allen, T. *Managing the Flow of Technology.* Boston: MIT Press, 1977.

[2] Brassington, F., and S. Pettitt. *Principles of Marketing.* London: Prentice Hall, 2006.

[3] Camner, G., E. Sjöblom, and C. Pulver. *What makes a Successful Mobile Money Implementation?* Retrieved November 15, 2009, from Mobile Money for the Unbanked, (2009): http://www.gsmworld.com/documents/m-pesa_case_study.pdf

[4] Chell, E. "Social Enterprise and Entrepreneurship; Towards a Convergent Theory of the Entrepreneurial Process." *International Small Business Journal* 25, no. 1 (2007): 5–26.

[5] Conway, G., and J. Waage. *Science and Innovation for Development.* London: Collaborative on Development Sciences (UKCDS), 2010.

[6] Development, W.C. *Our Common Future.* Oxford: Oxford University Press, 1987.

[7] Fyvie, C., and A. Ager. "NGOs and Innovation: Organizational Character-istics and Constraints in Development Assistance Work in The Gambia." *World Development* 27, no. 8, (1999): 1383–1395.

[8] Heeks, R. "Where next for ICTs and International Development?" In *ICTs for Development, Improving Policy Coherence.* OECD, (2009).

[9] Martin, F., and M. Thompson. *Social Enterprise, Developing Sustainable Businesses.* Hampshire: Palgrave Macmillan, 2010.

[10] O'Sullivan, D., and L. Dooley. *Applying Innovation.* London: Sage, 2009.

[11] Porter, M. *Competitive Strategy: Techniques for Analysing Industries and Competitors.* New York: Free Press, 1980.

[12] Price, M. *Social Enterprise, What it is and Why it Matters.* Wales: Fflan Ltd., 2009.

[13] Rothwell, R. "Successful Industrial Innovation: Critical Success Factors for the 1990s." *R & D Management* 22, no. 3 (1992): 221–239.

[14] Silverman, D. *Doing Qualitative Research.* London: Sage, 2010.

[15] Stake, R. "Case studies." In *Handbook of qualitative research*, edited by N. Denzin and Y. Lincoln. 2nd ed, 435–454. Thousand Oaks, CA: Sage, 2000.

[16] UNDP. *Human Development Report.* Oxford: Oxford University Press, 2001.

[17] Wheelwright, S., and K. Clark. "Creating project plans to focus product development." *Harvard Business Review*, Sept–Oct (1997).

References

[1] Aker, J.C., and I.M. Mbiti. *Africa Calling. Can Mobile Phone Make a Miracle?* Retrieved June 10, 2010, from Boston Review, (2010, March/April). http://bostonreview.net/BR35.2/aker_mbiti.php

[2] Bolton, B., and J. Thompson. *Entrepreneurs: Talent, Temperament, Technique.* Oxford: Butterworth- Heinemann, 2000.

[3] Bessant, J., and J. Tidd. *Innovation and Entrepreneurship.* Chichester: Wiley, 2009.

[4] Brundtland Report *('Our Common Future')* (1987): Oxford University Press.

[5] Burns, P. *Entrepreneurship and Small Business.* Hampshire: Palgrave Macmillan, 2007.

[6] Charitonenko, S., and D. de Silva. *Commercialization of Microfinance, Sri Lanka.* Manila: Asian Development Bank, 2002.

[7] Cooper, R. "Doing it Right: Winning with New Products." *Ivey Business Journal* 64, no. 6: (2000): 1–7.

[8] Drucker, P.F. *Innovation and Entrepreneurship.* Oxford: Elsevier, 1985.

[9] Fusion (2010): Social Impact Assessment in an ICT4D Intervention Using Theory of Change. *Discussion Paper.* http://enovation4d.blogspot.com/, (page live at 10.02.2012).

[10] Kutsoati, E., and S. Bartlett. (2008): *Texting TradeNet: What's the Price of Soya Beans?* Modern Ghana website (www.modernghana.com/news/181583/1/texting-tradenet-whats-the-price-of-soya-beans.html) (page live at 10.02.2012)

[11] Practical Action. (2010, January 12). *Defining appropriate technology (Frequently Asked Questions).* Retrieved June 22, 2010, from Practical Action: http://practicalaction.org/about-us/faq.

[12] Prahalad, C.K. *The Fortune at the Bottom of the Pyramid; Eradicating Poverty Through Profit.* New Jersey: Wharton School Publishing, 2004.

[13] Premasiri, P.D. Sri Lanka and the Sarvodaya Model viewed 31 May 2005, http://ignca.nic.in/cd_05012.htm, 1997.

[14] Schumacher, E. *Small is Beautiful: A Study of Economics as if People Mattered.* London: Vintage Book, 1973.

[15] Stamm, B.V. *Managing Innovation, Design and Creativity.* Chichester: Wiley, 2008.

[16] Tidd, J., and J. Bessant. *Managing Innovation. Integrating Technological, Market and Organizational Change.* Chichester: Wiley, 2009.

[17] World Bank (2006)

[18] Zainudeen, A., T. Iqbal, and R. Samarajiva. "Who's Got the Phone? Gender and the Use of the Telephone at the Bottom of the Pyramid." *New Media & Society* 12, no. 4 (2010): 549–566.

[19] Zainudeen, A., N. Sivapragasam, H. de Silva, T. Iqbal, and D. Ratnadiwakara. *Teleuse at the Bottom of the Pyramid: Findings from a Five-Country Study.* Colombo, Sri Lanka, 2007.

4

Conflicting Interests in User Requirements for Customization and Personalization

J. K. Sørensen

Center for Communication, Media and Information Technologies at
Department of Electronic Systems, Aalborg University, Copenhagen

4.1 Introduction

The term "user requirements" is today a well-established expression, and the description of the user requirements has become a common-sense element in every ICT development project. When something becomes common-sense tensions, conflicts or dilemmas which were visible as long as the term still was young tend to disappear, and the concept becomes often a taken-for-granted instrument to gain or produce something else, in this case ICT systems, software, and services. The instrument becomes invisible until it breaks or does not work any longer. If one however wants to analyze the now overlooked tensions, conflicts, or dilemmas in the process of identifying, formulating, or specifying user requirements, it is obviously most interesting to look at situations where the instrument breaks down. In this chapter, I discuss the two elements in "user requirements," namely "user" and "requirement," in relation to two concepts that have gained popularity in ICT services, particularly mobile services, namely "customization" and "personalization." This discussion takes its starting point in a theoretical approach but is in the second part of the chapter applied on a case study of the development of user requirements for a customizable web page with many stakeholders.

4.2 Who Requires What?

In ICT development projects, the process of formulating or identifying user requirements is often seen just as a short phase or step in a process leading to the release of the software or launch of the web service. Since most ICT projects are framed by time constrains, often this phase is kept as short as possible, reducing it economically to a "need to know" for the developers/programmers. Although that almost every project leader or programmer would state that the software or service should be "easy to use" and that the users are being thought of during the development process, the reality of ICT project management is often that the specification of user requirements is seen more as a construction needed to frame the programming work than a description of real users' requirements. The involvement of real users is time-consuming, and the users' views and opinions may change the planned agenda or objective of the project. The requirements specification serves as a blueprint for structuring the programming work, not as a description of the customers of the service. The user requirements thus become an abstraction or simplification of reality, a hypothesis about the users and their requirements. One could easily argue for the benefits of this narrow description of user requirements: In order to reduce the diversity of the real world down to a size which is manageable in an ICT development project, a large number of rare or seldom use-situations must be ignored when the user requirements are written. The reduction of the real-world diversity can be accomplished since real-world users can adapt to the systems by being trained to use the system. In some cases where no alternatives are provided to obtain the goal, users are even forced to use the software. The term "user requirements" may thus ironically rather be understood as "what the ICT system (or the actually the programmers) requires of the user," than as an expression of what the user claims or requires from the system.

These two perspectives of requirements could be described with the metaphor of texts, authors, and readers. Grint and Woolgar (1997) introduce the view that the software can be seen as a text written by an author (the programmer) and the use of it as a process of reading this text. Based on this view, we could say that there are different degrees of freedom to interpret the "text" depending on the system architecture and interface of the system; the text is open in different ways for interpretation, cf. Eco's (1989) concept of "the open work." The purpose of user requirements is thus to help programmers write the text while trying to anticipate as many ways of reading/interpreting the text as possible. The construction of the user requirements is based on the programmer's abilities to imagine the different attempts of reading the text/using the system. Some readings—or uses—of the system are preferred by the author, some are unwanted or even seen as malicious. The user requirements serve thus as the programmers' map of wanted and unwanted "readings" or paths through the system.

On a practical level, this "text reading" metaphor has however some drawbacks: It ignores that the person using the ICT system in most cases has a goal that lies beyond interacting with the system. If I interact with a ticket-vending machine, it is in most cases likely that my goal not is to explore the different readings of the "ticket-vending machine text," but that I want to travel with the train without risking to get punished for not having a ticket. If I type in an email or SMS again my goal is not to enjoy reading the "text," but to ensure that somebody else will get my message. I want the tool—the ICT system—to be as transparent as possible, except in the situation where I am interested in experiencing and reflecting on the "user experience" of the product or service (cf.: Wright and McCarthy, 2010). The responsiveness of the ICT tool in relation to my goals defines my personal agency, my individual freedom. The reading metaphor can thus be criticized for being centered on the author's perception of the world and understanding of user goals, not on the users'. The real-world users may have many different and very specific expectations of how the system-text should be "read," but they are constrained by the "author's" perception of and assumptions about users. The purpose of the user requirements specification becomes thus sadly not to depict users' requirements but to foresee, describe, and limit these expected readings, avoiding interpretations of the text which lead astray from the programmers' intentions.

4.3 Who is "The User"?

Historically, many attempts have been carried out to represent the user's interest in the ICT development. As in many other cases where professional planners (e.g., city planners, architects, or computer engineers) envision ideas and make decisions that influence "ordinary people's" (users', citizens', customers') life, tensions emerge. This is particularly the case when the consequences of the decisions are far reaching (Rittel and Webber, 1973). Many approaches, methods, and techniques have subsequently been suggested to counter these tensions. In the field of ICT development, it has thus since the emergence of participatory design (Ehn, 1988) been seen as good craftsmanship to include the users (as imagined or real persons) during the ICT development project (Buur and Bødker, 2000; Buur and Bagger, 1999).

There are however problems in user involvement: Who is the user? How can interaction designers, programmers, and developers know whether their assumptions about the users are true? Often, the user is an abstraction and simplification of the diversity of the real people who are likely to use the software. This simplification is not only a result of the need of freezing the list of possible features which the one or other user could wish for, but also a result of the economy of scale in software development: Generic pieces of code is being reused to lower development costs and decrease time spent on the project, and software or services are often sold or offered on a mass market. In order to be economically feasible, user requirements must be shaped by the resources available, meaning a less detailed description than the real-world users. The reduced description of the user has however also internal advantages for the design/development team in terms of creating shared visions for the product to be designed. Discussing the gap between real-world users— "people"—and the idealized "user" being applied in product development and marketing, McHardy (2009) suggests that the latter is essentially a construction that is based on the specific needs and visions of, for example, a development team, that users in this context are "creative interpretations of people from a situated point of view" (McHardy, 2009: 192). The "user" is thus a makeshift rhetoric device created by the development team or the marketing people rather than a real person in flesh and blood. The construction itself is shaped by a number of factors, among others the economic potential in the project, the wider strategic perspectives, the available technology, the organizational dependency, and the expected success on the market. Consequently, if little time is used in the "user requirement phase" and if interaction designers and user specialists often feel ignored in ICT projects, it is not a result of ignorance but of the (political) economy of the ICT development work. Still, however, this idealized user is being confronted by the real-world people when the software or service is launched.

4.4 Customization and Personalization

A typical approach to resolve the tension between the idealized and real user is to introduce customization or personalization in the system (MacLean et al., 1990). This introduces some flexibility between the programmers' assumptions about users and their real needs. However, when we look at the origins of customization and personalization, we find that the flexibility only exists on a normative level. Historically seen, in the world of physical goods, the term "customization" points at the relationship between a supplier/merchant and the customer and evokes the old proverb of the "customer king." The opportunity for customization is however ironically created by the breakdown of the "customer king" tale: The emergence of industrial production meant that the close and personal relationship, which allowed the merchant to honor the customer, has disappeared (Schipper, 2002). Customization is thus an industrialized approach to imitate some of the elements of the old customer–provider relationship, but within the framework of large-scale industrial production. Thus, customization is in the context of industrial production also referred to as "mass customization" (Victor and Boynton, 1998).

While the production and distribution of physical goods imply certain limitations to customization, there are wider possibilities in the world of ICT services. Individual modifications of the display or behavior—for example, to change the background color or picture, the structure of the menu, the shortcuts, the font size and font type—are mostly only a matter of changing parameters in the code. To a certain extent, one could claim that customization is an inherent quality of computers. Formally, customization of ICT services is thus rather a matter of granting access rights to change the software code. Ideally, the programmable ICT device would provide ultimate freedom and individuality to the users. In reality, this is not the case. The potential of the programmable nature of software code is only granted in small fractions to the end users, for example, as customizable interfaces. Ironically, tools for customization must to be designed and provided to the users. The ICT developers must again imagine a fictitious user for this purpose, again risking not to cover the gap between the real world and their own assumptions about the user.

The term "personalization" shifts the focus from the commercial setting of being a customer as in "customization," to normatively focus on the individual person. This promise is however also—as in the case of customization—difficult to fulfill. In the ICT context, the focus on the person is again typically from the perspective of the provider: The personalization literature (Fan and Poole, 2006) discerns between implicit and explicit personalization. Where the latter assumes an active user, like in the case of customization, the claim in implicit personalization is often that automated recommendation will make life easier for the user. Again, the claim is that the user requirements express the requirements of the users.

4.5 Personalized PSB News

To illustrate the tensions embedded between concepts "users," "require-ments," "customization," and "personalization," it is helpful to analyze a case where customization was implemented in order to meet expected—or constructed—user requirements. Furthermore, if we focus on a case where different stakeholders inside and outside of the organization have different perceptions of what users would require, tensions are further emphasized. Finally, if we choose a case where normatively every citizen in a nation-state is a potential user, the tensions are at their peak. Many private and public ICT services could fulfill these criteria, for example, self-service ICT solutions provided to citizens by tax authorities or online banking provided by private companies. An extreme case is however personalized versions of news web pages, provided, for example, by news papers or broadcasters: The news item, a product produced with a large unspecified audience in mind, is exposed to individual users based on some kind of individual filtering. A tension emerges immediately between on the one side the news producers' economy of scale and the interests in agenda setting, and on the other side the promise of customer sovereignty, individualism, and improvement of users' attention economy embedded in the personalization/customization concept, cf. Sørensen (2011). The personalization strategies of US and UK mainly commercial news publishers have been analyzed by Thurman (2011) and Thurman and Schifferes (2012), whereas the personalization strategies of three North European public service broadcasters (PSBs) have been analyzed by Sørensen (2011, 2013). It is one of the three PSBs, namely DR from Denmark, which will be analyzed in the following. The other cases show however similar signs of tensions regarding the shaping of the personalized news service (cf. Thurman, 2011; Sørensen, 2011, 2013).

North European public service broadcasters (PSBs) are interesting as case study for personalized news, since they have many different very users: Normatively, the whole population is the target. This obligation is both due to the history of North European public service broadcasting where the PSB organizations were—and still are—an important element in the cultural–political project of creating coherent societies and nation-states, and due to the financing of PSB, namely via a license fee or public financing. This political–economical construction of PSB leads to a dilemma: On the one hand, PSBs need to demonstrate that they are not undermining the business of commercial media; on the other hand, they need to demonstrate that they are appreciated by a large portion of the license fee payers or tax payers. This leads to a dilemma in the programming policies: Ideally large audiences should be served with content that is very distinctive and not catered for by the commercial media, cf. Nissen (2006). In praxis, PSB organizations need to find a balance between producing or offering content that is very unique for the PSBs, but possibly not very popular, and content that is very popular but possibly not very unique. Particularly the online activities of PSBs have lead to renewed discussions of the public service obligation: To which extent should the PSB activities be defined by and restricted to "broadcasting"? Or should PSB activities be independent of the media in which they take place (e.g., broadcast, web)? This question has lead to a very large body of literature and debate. The sheer volume of this prevents us from a detailed account of the different positions. The media political tensions around the PSB online activities are however an important background for the case study presented below. See for example ACT (2004), Graf (2004), Jakubowicz (2007), Moe (2008), and Löblich, (2011) for an extensive discussion.

4.6 The Case Study

Operating in a competitive media landscape, public service broadcasters are forced to position themselves also in respect to the ways media content is presented online. One strategy is to adopt styles and organizing principles that already have proven successful among users, or show clear signs of that. Of particular interest here are web pages that can be configured by the user to display specific content: customizable web pages. As customizable content aggregator web pages such as Netvibes and iGoogle were launched around year 2005, PSB web editors were not late to recognize the potential in offering PSB web users the opportunity to customize the PSB web pages. Sørensen (2011) presents an overview of North European public service broadcasters' web customization projects, as well as an analysis of three cases: "Mit DR" [My DR] from the Danish Broadcasting Corporation "DR," Denmark; "Mein WDR" [My WDR] from WestdeutscherRundfunk "WDR," Germany; and the customizable front page of BBC.co.uk, provided by British Broadcasting Corporation, UK. My analysis of these projects provides the background and inspiration for this chapter. In the following, I will however focus on one case, "Mit DR," which also is analyzed by Sørensen (2013).

The "Mit DR" project was analyzed through nine in-depth interviews over a period of three and a half years with the editors and project leaders directly involved in realizing the project. Furthermore, six interviews were conducted with other editors from the DR online department. Finally, design documents, interfaces, and functionalities, as well as media political documents, were analyzed to complete the picture and position the editors' intentions and ideas in both a media political context and an internal organizational context, and in relation to the actual outcome of the project: the customizable interface. This approach was chosen to identify drivers and barriers for the project, as well as for the analysis of it as a case of a public service media in relation to the media political discussion presented above. In the following, the interplay between these different drivers and barriers shall be presented. The following analysis indicates a very marginal role for the user, while other stakeholders have a larger influence on the outcome of the project.

The work leading to the "Mit DR" customizable web page was initiated on December 2006 as one member of the chief editorial board of DR online was assigned with the task to develop a concept for a personalized DR.dk Web site. The impulse came both from the commercial developments in social network services, like at that time quite small Facebook, and from the necessity of launching DR.dk's, at that time quite popular youth community site "SKUM." The idea of a customizable general web page for all DR.dk users that would allow them to pick and see only the content that would interest them was thus initially a spin-off. The two projects were initiated simultaneously in spring 2007 as a joint cooperation, but after six months they were separated into different projects. The re-launch of the youth community site was later abandoned, and the site closed in 2011. The "Mit DR" was released in an unstable beta-version in June 2008 and was re-launched in May 2009 based on another technology. In November 2009, an interface for presenting new users to the customization process was introduced: editor's selection of content was bundled in seven so-called start packages. In December 2009, the beta-phase ended. In January 2011, DR online department stopped maintaining the site, and in January 2014, the site was taken off-line.

According to interviews with the editors, an important reason for initiating the "Mit DR" project was the, at that time, very little layout flexibility of the DR.dk front page. The page was divided into fixed-size boxes, each managed by a content-producing department in DR. By offering a customizable web page, users would now themselves prioritize the content. This idea echoed a shift in the overall approach in the DR organization where listeners, viewers, and users were depicted and described by editors as "customers." A customizable web interface—"Mit DR"—as well as different on-demand services symbolized this shift.

When we look closer at the different factors driving the project, we can identify many stakeholders. The end users—the visitors of DR.dk—are just one. As it will be demonstrated in the following, the role of the end users, as well as their potential requirements, did ironically not play a big role, although they were the official target and objective of the project. Instead, we can observe many different and conflicting requirements, emanating from the other stakeholders.

Not only are PSB strategies defined by the media political conditions, but internally in the organizations resources are to be distributed between departments with different objectives and strategies. Departments responsible for content production have other interests than those responsible for marketing and channels regarding the presentation and exposure of the content. Departments that produce online services, like web pages, have other interests than those producing content intended for broadcast, stream, or on-demand, cf.: Sørensen (2011). Finally, a challenge in defining user requirements is that PSB organizations, like other publishers, both aim to act as agenda-setters for societal debates and political issues, and at the same time attempt to fulfill listeners'/viewers'/users' wishes and unexpressed desires for content. The composition of the media offerings is thus often determined by a combination of what editors think their audiences would be interested in and a desire to introduce content not already known or demanded by the audiences. The "user requirements" for the content—if one can use this term for the construction of audiences in traditional mass media scheduling (cf.: Ang, 1991)—are thus defined by both the actual exposure of a specific program (e.g., for TV, the "share") and by the editor and or media organization's interests in drawing attention to specific content. As we will see in the following, the user requirements for the web and mobile services delivering the PSB content are however also depending on the different interests that shape the PSB activities.

The interviews conducted with the editors and project managers involved in the "Mit DR" project show that the initial ideas of providing a high degree of user freedom gradually faded during the project. The "Mit DR" project originally aimed at giving DR.dk's users a highly customizable page that, like Netvibes, would allow the user to collect media content from many different sources, for example, Youtube and LastFM at a web page hosted at DR.dk. The DR editors' assumption was that this would increase the time spent on DR.dk, and at the same time nurture social communication among the DR users: The personal and customizable "Mit DR" pages were planned to be accessible to other DR users and thus serve as a kind of viral marketing for DR content. This vision—or "user requirement"—was not based on any interviews or involvement of users, but on the editor's observations of other web services, like at that time booming "Facebook" and Netvibes. This vision downplayed DR's agenda-setting public service mission of presenting editorially selected content to users that they would not know, but might find interesting. Instead, the position was that DR's content had to compete as on a marketplace with other content at DR's own Internet site. The user requirement thus embedded an idea of customer sovereignty, which Poder described in an interview as fundamental for all web services.

As the implementation of "Mit DR" started, selected users were invited to a workshop to tell the new editor their ideas, habits, and expectations for a customizable DR.dk. This generated an initial list of possible features—the so-called widgets—of the customizable page. Soon, however, the development team had to prioritize, and chose to develop a generic RSS-feed reader that could display RSS-news feeds issued by the different content-producing departments in DR, as well as external feeds. Although other widgets were selected for production, such as an on-demand/streaming TV widget, a radio streaming widget, a "my area" widget, a "my program schedule" widget, an "archive/old programs" widget and widget for music reviews, only the generic RSS-feed reader and archive TV widget were released, as technical problems hampered the ambitions.

The lack of other types of widgets resulted in a limited variation in "Mit DR's" catalog of widgets, since only RSS-feeds could be shown. The limitation in types of widgets was however countered with a large selection of RSS-feeds. The collection of RSS-feeds at DR.dk thus counted 210 on January 31, 2010. This high number was achieved by including a large amount of RSS-feeds from external sources in the catalog. A total of 143 widgets/RSS-feeds (equal to 68.1%) were from external sources. They were selected by the "Mit DR" editor from the top-100 list of .dk news sites, and from international RSS-feeds, reflecting both the most popular ones as well as those covering areas where DR do not produce much content (e.g., lifestyle and sports).

Other decisions contradicted the claimed user requirement of customer sovereignty. To prioritize the production of widgets, DR's traditional mass media oriented segmenting tool that divided users into those interested in news, the nerds, the women, etc. was applied. This segmentation approach was also used for new users. New users were at a configuration page prompted to choose one out of seven different "packages": (1) news, (2) sports, (3) entertainment, (4) family, (5) life style, (6) nerds, and (7) "the overgrown teenagers" package ("Drengerøvspakken"). A text explained to new users that the packages could be modified, but according to the DR editors, users did only this in a very small scale. The user requirement assumption behind the "start packages" thus contradicted the original visions with the project that stressed customer sovereignty and free choice.

The user requirements appeared thus to be contradictory and changing during the project. One explanation can be that the actual users of the system were not very well represented in the project. Except a user workshop, a usability test, and a weblog where the "Mit DR" project leader informed about design decisions, users were not involved.

A number of different stakeholders can be identified, some internal, some external. Space does not allow a substantial description of each of these internal and external stakeholders' interests in the "Mit DR" project (see Sørensen, 2011 for comprehensive overview), but we can observe that the actual end users only are one among many. Internally in DR, stakeholders can be found at all levels in the organization: (1) DR's chief editorial board, (2) DR.dk's Editor in Chief, (3) the technical department TES, (4) the marketing department, (5) the content-producing departments, (6) the Interactive Productions department (programming the "Mit DR"), (7) the youth web department (initially assigned with the programming task), and (8) the different editors. Each of these internal stakeholders was driven by each of their particular interests in the system.

Among external stakeholders were other media content producers in Denmark, for example, RSS-feed providers, as well as Microsoft which delivered the SharePoint server solution initially used, later abandoned. Finally, since DR is a public service media organization, the discussion of possible anti-competitive effects of DR's Internet activities also influenced the project. The "Public Service contract" 2007–2010 between DR and the Danish ministry of Culture (2006), as well as the measurements made to evaluate the "public service value" of DR's services—the so-called værditest (Kulturministeriet, 2007)—which follows a European trend of submitting particularly online services from public service broadcasters for an evaluation of their contribution to public value compared with their potential bias of the forces in the media and online market, for example, "DreiStufen Test" in Germany (§11 in 12. Rundfunkänderungsstaatsvertrag, 2009), and the public value test in the UK (BBC, 2012). The general political debate on DR's remits could also be seen as influencing the "Mit DR" project.

The discussion of the different interests and the possible neglecting of the actual users in shaping the service could be abandoned if the "Mit DR" had become a success. But, although the editors expected 200.000 users of the single-sign-on system used for "Mit DR," only 12.000 signed up for the service. Furthermore, throughout the period December 2009–October 2010, where "Mit DR" was fully developed, it only attracted 0.09% of the total number of visits at DR.dk, and only 0.031 percent of the page views. As of January 2014, the service has been taken off-line. The "Mit DR" did thus not fulfill the user requirements—or more precisely—the users required something else.

4.7 Conclusion

The case of "Mit DR" is obviously not the only case of ICT development where the producers were mistaken about user requirements, or where user requirements were constructed rather to create a compromise between internal and external stakeholders, than to reflect actual user needs. What is worth observing (with McHardy, 2009) is that those proposing, shaping, and negotiating the service always construct users. The service is thus typically more a statement to the world than a result of user requirements. I thus want to argue that the term "user requirements" is misleading, since it obscures the role of all the other stakeholders. This becomes particularly visible in a multi-stakeholder context, such as in the case of "Mit DR." One could speculate whether this also holds for other big institutions. The irony of user requirements of customizable services, or services that use customization, is that customization exactly is used with the objective of giving users more freedom to shape the application. The customizable features in a service could thus be seen as a sign of the design team's wish to delegate decisions to users, and as a wish to cover broader user requirements. However, although customization appears as the solution for assumed diverging user requirements, it is often just obscuring the problem. The conflicting interests between the different stakeholders are being obscured by delegating a few, not very important, decisions to users while the important design decisions are taken on the basis of user requirements that echo stakeholder interests.

References

[1] Ang, I. *Desperately Seeking the Audience*. London, New York: Routledge, 1991.
[2] ACT. (2004). Safeguarding the Future of the European Audiovisual Market—A White Paper on the Financing and Regulation of Publicly Funded Broadcasters. Association of Commercial Television in Europe, Association Européenne des Radios, European Publishers Council. Retrieved from http://www.epceurope.org/presscentre/archive/safe guarding_audiovisual_market_300304.pdf
[3] BBC. *BBC Trust Assessment Processes Guidance Document*, 2012.
[4] Buur, J., and K. Bagger. Replacing Usability Testing with User Dialogue. *Communications of the ACM* 42, no. 5, (1999): 63–66. doi:10.1145/301353.301417
[5] Buur, J., and S. Bødker. "From Usability Lab to "Design Collaboratorium": Reframing Usability Practice." In *DIS '00—Conference on*

Designing Interactive Systems: Processes, Practices, Methods, and Techniques, edited by D. Boyarski and W. A. Kellogg, pp. 297–307. New York, NY: ACM Press, 2000. doi:10.1145/347642.347768

[6] Bødker, K., and J. Simonsen. *Participatory IT DESIGN. Designing for Business and Workplace Realities*. Cambridge, MA: MIT Press, 2004.

[7] DR og Kulturministeren. (2006). Public service kontrakt mellem DR og kulturministeren for perioden 1. januar 2007 til 31. december 2010. Kulturministeriet. Retrieved from http://www.kum.dk/graphics/kum/downloads/Kulturomraader/Radio_og_TV/Public Service kontrakt 2007/Publicservicekontraktennyudgave.pdf

[8] Eco, U. *The Open Work*. Cambridge, MA.: Harvard University Press, 1989.

[9] Ehn, P. *Work-Oriented Design of Computer Artifacts*. Stockholm: Arbejdslivscentrum, 1988.

[10] Fan, H., and M.S. Poole. "What Is Personalization? Perspectives on the Design and Implementation of Personalization in Information Systems." *Journal of Organizational Computing and Electronic Commerce* 16, no. 3–4, (2006): 179–202. doi:10.1080/10919392.2006.9681199

[11] Graf, P. (2004): Independent Review of BBC Online. London: DCMS/Reckon. Retrieved from http://go.reckon.co.uk/2jot

[12] Gregory, J. "Scandinavian Approaches to Participatory Design." *International Journal of Engineering Education* 19, no. 1 (2003): 62–74.

[13] Grint, K., and S. Woolgar. *The Machine at Work*. Cambridge: Polity, 1997.

[14] Jakubowicz, K. "Public Service Broadcasting in the 21st Century. What Chance for a New Beginning?" In *From Public Service Broadcasting to Public Service Media,* edited by G.F. Lowe and J. Bardoel, 29–49. Kungälv: Nordicom, Göteborg Universitet, 2007.

[15] Kulturministeriet. (2007): Værditest af nye public service-tjenester. Bilag til Public Service kontrakt 2007–2010. Retrieved from http://www.kum.dk/graphics/kum/downloads/Kulturomraader/Radio_og_TV/Public Service kontrakt 2007/Bilag 2 Vaerditest af de nye public service programmer.pdf

[16] Löblich, M. "The Battle for "Expansion" of Public Service Broadcasting on the Internet. The Press Coverage of the 12th Amendment of the Interstate Treaty on Broadcasting and Telemedia in Germany." *International Journal of Media and Cultural Politics* 8, no. 1, (2012): 87–104. doi:10.1386/macp.8.1.87_1

[17] Nissen, C.S. "No Public Service Without Both Public and Service—Content Provision Between the Scylla of populism and the Charybdis of Elitism. In *Making a Difference: Public Service Broadcasting in the European Media Landscape*, 65–82. Eastleigh: John Libbey Publishing, 2006.

[18] McHardy, J. "Make-Shift Ciopfaspovs: an Exploration of Users in Design. In *Multiple Ways to Design Research*, edited by G. Anceschi, 181–194. 2009.

[19] MacLean, A., K. Carter, L. Lövstrand, and T. Moran. "User-Tailorable Systems: Pressing the Issues with Buttons." In *Proceedings of the SIGCHI Conference on Human Factors in Computing Systems Empowering People—CHI '90*, 175–182. New York, NY: ACM Press, 1990. doi:10.1145/97243.97271

[20] Madurapperuma, A.P., W.A. Gray, and N.J. Fiddian. "Customisable Graphical Interfaces to Database Systems: A Meta-Programming Based Approach." In *Database and Expert Systems Applications. 8th International Conference, DEXA '97. Proceedings*, 318–323. IEEE Computer Society, 1997. doi:10.1109/DEXA.1997.617300

[21] Moe, H. "Public service broadcasting påinternett? En komparativ analyse." In *Public service i netværkssamfundet*, edited by F. Mortensen, 67–101. København: Forlaget Samfundslitteratur, 2008.

[22] Rittel, H.W., and M.M. Webber. "Dilemmas in a General Theory of Planning." *Policy Sciences* 4, no. 2 (1973): 155–169. doi: 10.1007/BF01405730

[23] Schipper, F. "The relevance of Horkheimer's view of the customer." *European Journal of Marketing* 36, no. 1/2 (2002): 23–35. doi:10.1108/03090560210412683

[24] Syam, N.B., and N. Kumar. "On Customized Goods, Standard Goods, and Competition." *Marketing Science* 25, no. 5 (2006): 525–537. doi:10.1287/mksc.1060.0199

[25] Sørensen, J.K. *The Paradox of Personalisation: Public Service Broadcasters' Approaches to Media Personalisation Technologies*. (2011) University of Southern Denmark (SDU). Retrieved from http://nordicom. statsbiblioteket.dk/ncom/da/publications/the-paradox-of-personalisation %28e0d6ae05-2a6d-453e-8a1b-72df8cf21457%29.html

[26] Sørensen, J.K. "PSB Goes Personal: The Failure of Personalised PSB Web Pages." *MedieKultur* 29, no. 55 (2013): 43–71. Retrieved from http://ojs.statsbiblioteket.dk/index.php/mediekultur/article/view/7993

[27] Thurman, N. "Making "The Daily Me": Technology, Economics and Habit in the Mainstream Assimilation of Personalized News." *Journalism* 12, no. 4 (2011): 395–415. doi:10.1177/1464884910388228

[28] Thurman, N., and S. Schifferes. "The Future of Personalization at News Websites." *Journalism Studies* 13, no. 5–6 (2012): 775–790. doi: 10.1080/1461670X.2012.664341

[29] Victor, B., and A.C. Boynton. *Invented Here*. Boston, MA: Harvard Business School Press, 1998.

[30] Woolgar, S. "Configuring the User: The Case of Usability Trails." In *A Sociology of Monsters: Essays on Power, Technology and Domination*, edited by J. Law, 57–102. New York: Routledge, 1991.

[31] Wright, P., and J. McCarthy. "Experience-Centered Design: Designers, Users, and Communities in Dialogue." *Synthesis Lectures on Human-Centered Informatics* 3, no. 1, (2010): 1–123. doi:10.2200/S00229ED1V 01Y201003HCI009

[32] Zwölfter Staatsvertrag zur Änderung rundfunkrechtlicher Staatsverträge (ZwölfterRundfunkänderungsstaatsvertrag) (2009): Das Land Baden-Württemberg, der Freistaat Bayern, das Land Berlin, das Land Brandenburg, die Freie Hansestadt Bremen, die Freie und Hansestadt Hamburg, das Land Hessen, das Land Mecklenburg-Vorpommern, das Land Niedersachsen, das Land Nordrhein-Westfalen. Retrieved from http://www.ard.de/download/138948/index.pdf

5

Security and Usability

Jing Chen[1]**, Marcus Wong**[2] **and Lijia Zhang**[3]

[1]Huawei Techonlogies, Shanghai, China
[2]Huawei North America R&D Center, Bridgewater, NJ, USA
[3]Huawei Technologies, Beijing, China

5.1 What is Usable Security?

Security for every service and application we depend on and use every day is turning into a major challenge for all of us, not just the designers, the architects, the developers, and implementers alike, but especially so for the users. This is especially true for the new services and applications built on the new architecture for the future. After all, a system is as good as the users that use them. As users, we have the firsthand experience of the system. If a system has the greatest security features and yet is not convenient or usable for the users that it is intended, not only would the security of the system fail, but the overall system fails as well. It is important to build a secure and reliable system. It is equally important to build a secure and reliable system that is usable at the same time.

What is usable security and can usability and security go together? Usable security is about making security as transparent and understandable as possible and yet visible at the same time. Usable security gives the end users security controls so that they can understand and control privacy in the dynamic, pervasive computer environment of the future. Users do not interact in isolation. Instead, they are integral part of a much larger community of users that interact, communicate, and share their everyday experiences. In sharing these experiences, the users help system designers to enhance user-friendliness in positive ways. The smallest weakness of a system or the slightest dissatisfaction from a user will be magnified thousands of times. It will be put in the spotlight for everyone else to see. For applications and services with greater visibility, any issues related to either usability and security may also turn into front-page news. On the other hand, usability and security can go together to enhance the user experience and maintain acceptable level of security at the same time.

5.2 Background

5.2.1 Usability Status

5.2.1.1 Usability in mobile communication

With the advance in mobile communication systems, the user experience has become a way of introducing the users to new products and services. Many companies have succeeded in introducing to the end users easy-to-use and easy-to-customize devices. The days of the ponderous brick-sized cell phones are over, and yet there are more and more features built into the end user devices than ever. Not only are they smaller, lighter, and more esthetic, but they are more intuitive to use and last longer. For example, for traditional candy-bar handsets with a keypad, even a blind person can make a phone call without assistance. The keypad layout, and the "send" and "end" buttons have almost become universally acceptable in strategic positions of the handset for both convenience and ease of use. They are almost instantly recognized as to the location of the buttons on the handset. All in all, the industry has gone through a revolution of usability that not only benefited the end users, but they have become the norm for all.

5.2.1.2 Usability in other systems

As more and more consumer electronics and other end systems are available to the users, the usability aspects are continuously being evolved for the better. For example, in the early days of automobile navigation systems, some vendors have been criticized as being "unintuitive" by requiring the user to confirm every command or the requiring going back to the main menu by backing up step by step. As the technology matures, the vendors have listened to customer feedback. They have learned to make things easier to use. Though more complex and with much more additional features than the first-generation systems, the newer versions of these navigation systems have become friendlier to users. When user frustration eases, the satisfaction level increases and usability has succeeded. In general, as with all end user equipment, usability has gotten better and better.

5.2.2 Security Status

5.2.2.1 Security of mobile communication system

With the development of mobile telecommunication, security becomes more and more important.

The first-generation mobile communication system hardly takes any security measures. The user (i.e., mobile terminal) sends its identity to the network in the form of plaintext. If the user identity sent can be mapped with the identity that the network stores, the user is authorized to access to the network.

Problem: Identity theft. The attackers pretend to be the legal users in order to access to the network.

The second-generation mobile communication system adopts private key-security mechanism. Some threats exist in the aspects of identity authentication and encryption algorithms.

Problems: The key used between subscriber identity module (SIM) and Authentication Center (AuC) can be easily eavesdropped, so that the attackers can obtain the key and clone the security context; the system only encrypts the message transmitted on the air interface (between the mobile station and the access network), so the attackers have the chance to intercept the messages which are transmitted in plaintext; it's difficult to detect if data are tampered in the transmission process because the system does not provide data integrity protection in the air interface.

Aiming at the problems above, the third-generation mobile communication system provides better security protection for the communication. With the evolution of 3G, the security functions are gradually improved and extended. The integrity protection and confidentiality protection are defined in Release 99 of the UMTS system. The protection to SS7 (Signaling system No. 7) in core network and in IP-based signaling is enhanced further in Release 4. The IMS security mechanism is added into Release 5. Generic authentication architecture (GAA), the security mechanism of Multimedia Broadcast Multicast Service (MBMS), and the security mechanism of UMTS interworking with WLAN are added into Release 6. In LTE/SAE system, even more care is taken in designing the security. Still, there are other issues.

Problems: Tracking UE, Identity theft, Unauthorized Access, DoS attacks, and so on.

5.2.2.2 Internet security

As the Internet becomes more mature and a part of our daily life, technologies used to provide secure services of the Internet under all circumstances become more crucial. Since network attack is very easy to deploy and has a very low cost, it has become the main reason of network instability and is the most challenging of the current network security.

Although a variety of attacks can be launched, they could be categorized into two domains: host-based attack and network-based attack. Although they are aiming at different domains, the most famous attacks that affect network security are as follows: denial of service (DoS), Web site defacement, virus and worm, data sniffing and spoofing, unauthorized access, and malicious code and Trojan.

DoS attack has the potential to be the most threatening one. It needs relatively low cost and causes huge damage in comparison with other attacks. DoS attack can cause victims, service providers, routers, and even the entire Internet to experience serious instability and performance degradation.

5.2.3 Security in the Future

Learning from the current status of security, we assume that the security mechanism in the future can do the following:

- provide security functions of mutual authentication
- provide integrity and confidentiality protection from end to end
- adopt more advanced cipher system
- be usable to the users (e.g., users can select the level of the security they want)
- adopt authentication based on biologic technology (e.g., a class of personal authentication techniques based on measuring innate cognitive abilities of the human brain)
- provide security event management (e.g., analyses network security event data in real time for threat management), and so on.

Based on the above requirements, the security in the future should have the function to resist the attacks and threats that exist in the system. However, it is not enough. Why and what other attributes should the security in the future have? Look at the example below, then you will readily understand.

There is a woman who lives in a big city, and like other citizens she has a healthy concern for safety. Her front door has a deadbolt, a latch, a chain, an extra deadbolt, sliders top and bottom, an intercom system, and a high-tech peephole. It can take two minutes to vet the pizza delivery guy and permit access for a pepperoni with extra cheese. But when she takes out the rubbish or goes to the curb to get the morning paper, more often than not she leaves the front door wide open. It is too much trouble to deal with all those latches and locks, codes, and keys. Good technology, but sometimes security can make things less secure.

The same is true when designing security with usability in mind in the future. Often, the more the security is technically complex, the more users will thwart the security. Besides the security, usability, which makes the security easy and acceptable to the users, and effectiveness, which enables users to achieve their goals at an appropriate level of security, should also be taken into account. In another word, we should adopt security with usability in mind.

Usable security is a service. It is an increasingly self-organizing auto-nomous service with a dynamically changing context. Multiple networks can provide usable security service. The same level of security is maintained without user intervention when users roam between two networks.

5.3 Stakeholders

Usable security involves a wide range of participants, which are called "stakeholders." And security is the process of keeping the risks of an infor-mation system at a level acceptable to all stakeholders. Since everyone has a stake in a chain of complex processes leading from a concept to an end product. These stakeholders may be both direct and indirect, for example, customers are certainly the stakeholders, but developers and related are also the stakeholders. As long as someone has a stake, that someone is a stakeholder.

In detail, stakeholders come in a variety of flavors: end users, corporate users, developers, system administrators, network operators, manufacturers, service providers, regulators, government agencies, certification authorities, insurers, and attackers.

5.3.1 Individual Users

Obviously, individual users including the end customers are on the very top of the stakeholder list, especially for usable security.

5.3.2 Corporate Users

Corporate users, whose organizational employees use future services and applications for corporate business, or even government and military users, who demand higher level of security than most are considered stakeholders.

5.3.3 Developers

Developers, such as system developers, application developers, and graphics designers who are integrated from the initial stages of inception to full product realization are stakeholders.

5.3.4 System Administrators

System administrators including those involved in network operations and network provisioning are stakeholders in that the day-to-day operations of all services and applications depend on them.

5.3.5 Network Operators

Network operators provide network access, and service access cannot be excluded. Network operators can be different operators; also, access technologies can be of many types. As end users employ different services through different accesses, heterogeneous networks security should be considered.

5.3.6 Manufacturers

Manufacturers provide the terminals and network equipment without which there would be no services and applications to run on. As usable security should be made acceptable to all stakeholders, including the terminals and network equipment, manufacturers need to produce more humanization and individualized human–computer interface terminals and network equipment.

5.3.7 Service Providers

Service providers such as banks, merchants, and everyone in the supply chain may not be part of the day-to-day operations, but they provide the flow of goods and services in a transparent way for the end users. So usable security for service provider means minimal support requirements.

5.3.8 Regulators

Laws and regulations lay the foundation for which all stakeholders are bound to abide by. Laws and regulations in different countries or regions may be different. Regulators draft out laws and regulations, which involve usable security, that may influence other stakeholders.

5.3.9 Government Agencies

Regulators in the government agencies that provide security services, police services, and information ministry are all stakeholders.

5.3.10 Certification Authorities

Certification authorities including trusted third-party services and broker services are stakeholders. They are essential part of systems that rely on certificate-based security.

5.3.11 Insurers

Insurers are the corporations who pay for the failure of the services and applications. When things go wrong, liabilities are involved and insurers including adjusters and fraud investigators become focal point of the stakeholders.

5.3.12 Attackers

Even attackers and hackers are stakeholders. What they do and what we do to prevent what they do is big part of reason why security measures are put in place. Ethical hacker provides good feedback to improve the security of the system, while malicious hackers and attackers are the ones that we need to keep out of the system.

There may be other stakeholders as well, all of which contribute to the success or the failure of the services and applications.

5.4 Likely Threats

Usable security for users must be secure and easy to use. But designing and creating services and applications with security and usability in mind is not an easy task. There may be many potential issues and problems that we face. In addition to this, designers of such a usable secure system also face many likely threats and attacks. Likely threats are considered the net negative impact of the exercise of vulnerability. In the following, likely threats and attacks are briefly presented.

5.4.1 Loss of Device

Careless when using or stolen by attackers may result in loss of device. Loss of device may lead to the loss of data, loss of history information, and loss of contacts in phonebooks.

5.4.2 Identity Theft

Compared to loss of device, loss of identity can be much worse. Identity theft yields more than just financial losses for the users and services providers; moreover, it may result in mistaken identity, which has potentially long-lasting legal ramifications.

5.4.3 DoS Attacks

The denial of service (DoS) attacks is more than just a nuisance; it creates unavailability for the legitimate users and takes away their ability to fully realize the benefits of the very service or application that they are entitled to. DoS attacks mainly consist of two types. The first type is about sending many packets to the server or the host, and the server cannot respond useful information because of busy dealing with this useless data packets. And the other type is disturbing the communication between the server and the host directly.

5.4.4 Unauthorized Use

Unauthorized use is another threat for usable security. It may include several types. And the main problem of unauthorized use is unauthorized access, which include stealing processing power, bandwidth, and subverting services, illegally sharing unprotected open Wi-Fi connections, bandwidth hijacking, and channel hijacking.

5.4.5 Malicious Code

Malicious code contains worm, virus, Trojan, and so on. Open source is a good thing which makes the Internet today so successful, but with that in mind there is also increased threats and attacks associated being open. Risk increases if handset or device becomes open (currently proprietary/closed), as these terminals or devices are becoming unintended carriers and infectors of malicious worms and viruses. For example, Java-enabled phones may become manageability of devices through network operator because of openness of the Internet. But openness also helps in promoting greater awareness that lead to greater responsibility for all of the stakeholders.

5.4.6 Surveillance

Legal intercept is an important legal requirement in almost every country. When surveillance is not observed properly under the laws of legal intercept, it may become stalking. Stalking not only is a serious threat to personal privacy, but also creates an intimidating environment for the users.

5.4.7 Unintended Disclosure

Unintended disclosure of data leakage and privacy may be the result of loss of personal control, for example, too much personalized marketing called "brain rape" that preys on naive and uneducated consumers for a small prize or token of nominal value. These over-seduction consumers may unintentionally and eagerly let their data or privacy guard down. It is a threat that needs to be taken into consideration.

5.4.8 Unwanted Persistence of Data

Unwanted persistence of data is another problem, which may be violating regulatory obligations of user privacy. Many Internet searching technologies like Google cache may preserve data for marketing or research purposes. If attackers steal this data, users' searching history and other information may be exposed to the attackers.

5.4.9 High Technology Crime/Harassment

As the development of Internet technology, high-tech crimes for business values or tricks like hackers will be a big problem to all stakeholders. These can be in the form of SMS bullying and happy slapping.

5.4.10 Tricking/Attacking Individual Users

Once attackers trick with individual users not the computers, the situation will become worse. According to designers, poor feedback or lack of feedback can make designing usable security difficult. Tricking/attacking individual users can be in the following forms:

- Spam
- Phishing
- Visual deception
- Using premium-rate services
- Social engineering

5.4.11 Insider Attacks

Disgruntled employees may be a big threat to the company's security. Insider attacker can easily render the system inoperable because he/she is more familiar with the system than other attackers.

5.4.12 Security as Business

As security is considered as business, then it would be also a problem because attackers can profit from users' threats or vulnerabilities. For example, virus protection and firewall product vendors will impose or aggrandize users' threats or vulnerabilities to sell their products. However, these products may be useless or sometimes cumbersome for the users.

5.4.13 Strong Security as a Problem

Usable security provides users with efficacious and understandable security. But it does not mean that more strong security provided more perfect for the users. Strong security can also be a problem for usable security. For example, encrypting too strong will make export difficult or make data irretrievable in the event of encryption key loss.

5.5 How to Secure

It's impossible to say users will achieve absolute security even though we can strike a good balance between perfect security and usable security in the future. The development of security technology is going with the amelioration of attackers, which is always in dynamic change. How to secure is an extensive problem to deal with. Treating the whole system as a cask, we just need to make every board long and strong enough in order to keep much more water in the cask. All of stakeholders are involved in security mechanism; beyond current security system, usable security provided with some new characteristics show the developing foreground of security. Herein we will describe the pivotal problem how to secure as follows in brief.

5.5.1 Trusted Components and Data Minimization

As the precondition to achieve usable security, the trusted components of the whole system include trusted computing platform, trusted processor, and trusted memory, also considered as base of all the security mechanism. Though nowadays depreciation of memory storage device makes end users take no account of the waste redundancy data but only consider what they really want to get, the system designers and developers still have to reduce redundancy to make data minimization. One should not ask for any data not necessary for transaction nor should he identify unless absolutely necessary such as use authorization mechanism. One should not record unless absolutely necessary.

Security gets circumvented if people are under pressure or for convenience (recall our earlier example of a lady getting newspaper), otherwise priority becomes low in large-scale complex economies, and redundancy may increase precarious ingredients or lead to unpredictable bugs that affect security. On the other hand, the necessary redundancy requested by some system mechanism and procedure is irrevocable.

5.5.2 Use Security Policies (Based on Risk Assessment)

Security policies based on risk assessment are properly used in most security system at present. The security classification for various applications requires dynamic management, which is mainly treated as proper security policies including three parts: preparation, prevention, and response. Preparation mainly includes following works: create usage policy statements, conduct risk analysis involving all stakeholders except the attacker, use risk analysis/development method that integrates security and usability in specification, and establish a security team structure. Prevention means approving security changes and monitoring security of network by multifarious methods. Response includes warning of security violations, restorations, and reviews from error or attack. In the future, risk assessment tends to be more intelligent along with the development of artificial intelligence. Maybe we can imitate the update of anti-virus software, and create a database of diverse security risk scenarios, in order to change the security policies of different applications timely in a dynamic mode. What's more, durative updating will make security classification more reasonable and believable.

5.5.3 Reliable "Delete"—Complete Destruction of Data

Another point called reliable "delete" meaning complete destruction of data, which is neglected by many users, therefore, may lead to loss as a result of the filching of attacker; for example, we have deleted the data on our hard disks, but in fact the data was not really deleted unless writing new data to overlay previous data, otherwise attacker could recover previous data by some methods. We need to take notice that pointer to data unavailable to OS (operation system) is not the same as destroying and avoid rewriting over data to prevent data reminisce. Defense in depth (several lines of defense) has become a common conception in information systems today. When we surf on the Internet, gateway, router, firewall, and anti-virus software help us avoid suffering from being attacked; without question we should never reckon on single line of defense, especially in usable security, where we must pay attention to layers of security (appropriate levels of security for risk), people defense, security technologies defense, and operations defense.

5.5.4 User Authentication and Mutual Authentication in Most/All Transactions

User authentication and subscription authentication are essential for any service and application. Authenticating users in addition to the devices alone is required to prevent the illegitimate users from accessing a legitimate user's services and applications. With the dramatic advance of biometric technologies, it provides increased security and yet is convenient and transparent to use. Many of the biometric technologies easily provide enough security to satisfy authentication requirements via two forms of authentication, namely identification and verification. They have once been thought as too complex to use and give too much false positives but are already being used again and again and many are even built into portable devices. These include fingerprinting, face recognition, retina scan, DNA, voice, even keystroke, and handwriting.

Whenever possible, use mutual authentication in most, if not all of the transactions to protect both the user and the server that provide the services or applications in which the user is accessing. Rogue servers and rogue networks do exist more often than most people think. Collaborations between user and network are essential in providing good security. We simply cannot rely on network alone because the networks are as secure as the users who use them. Encryption and message authentication is one of the best ways to protect data on devices and between data paths by protecting both signaling and identities. Message authentication combined with digital signature provides irrefutable evidence about the tractions. These tools are also an essential part of every secure system.

5.5.5 Better-Informed Users and Instant User Feedback

Usable security tries minimizing physical and mental workload of users. Therefore, instant user feedback must act as a supporter helping developers to improve the system constantly since making a security decision correctly is not easy without accumulative experience. Correct reuse, which may include elicitation, of concentrated security knowledge and functionality is helpful for security and usability and will make security more homogeneous and predictable. In a word, when something works, continue to use it, otherwise improve it. In addition, better-informed users help a usable security system to be more secure, better education means better awareness, better awareness means better users, better users mean better security. Come back to the trope, how much water a cask can hold is always decided by the shortest board—users are different with different education, different knowledge, and different experience—the devisers and developers of usable security should never make users become the shortest board in stakeholders. In other words, trying reducing the mistakes made by users requires better understanding of user's decision-making process and imperative decision when it's necessary. Users always prefer better-designed user interfaces because they do not want to know how the system works but how to use it, just like how advanced programming languages are far more popular than machine language and assembly language; besides, user or system actions to avoid or recover from security-related errors need to be part of reuse contract or interface of the components. In any case, users should not become the main reason leading to errors or hidden trouble of safety.

5.5.6 Better Understanding of User Decision-Making Process

How does better understanding of user decision-making process help in designing good security? It helps users to help us better by providing better-designed user interfaces. Good users provide good feedback. User or system actions to avoid or recover from security related–errors need to be part of reuse contract or interface of the components. A systematic development approach is just a good security practice. It not only provides good design by avoiding focusing on functionality alone, but we design with our audience in mind by thinking like the audience. Eliminate clutter and complexity. When usability and competitive pressure collide, be a user advocate by standing on the side of the user. A good development process alone does not make a good and secure system; a good development process has to be complemented by a good implementation. Key to good implementation is to minimize mistakes often made in implementations and test thoroughly, extensively, and exhaustively. Whenever user feedback is available, use it as a guide for constraints to make security decision correctly and easily. When something works, reuse it for security and usability by concentrating on security knowledge and functionality and making security more homogeneous and predictable.

5.5.7 Good Implementation Complements Good Development Process

To appraise the performance of a system, of course practicality is one of the most important parts that should be taken into account on production and probation period. We must test extensively and exhaustively to minimize the mistakes appearing in implementation before mass-producing, what's more, good implementation means not only less mistake, but also the satisfaction of users and perfect performance on what we really want to achieve, for usable security, that's usability known from other security system we are using nowadays. Here an exhaustive system log plays an indispensable part on development supporting, especially on the hand of operation management, flux percolation, bug correction, and attack alarm. When some parts work abnormally or even break down, the system administrator first checks the log to find out what's wrong; in addition, he may also enquire users, hereinbefore the importance of instant user feedback has been emphasized, which also indicates that good development process is determined by both comprehensive design conception of developers and requirement of users.

5.5.8 Systematic Development Process

The overview of usable security systematic development process can be described as follows. First, it is a good design based on the requirement of users, avoiding focusing on functionality alone, which means designers must pay more attention to your audiences and whom you think like your audiences in order to meet their opinions. Second, risk analysis must be carried out involving all stakeholders (minus the attacker), for risk analysis/development method integrates security with usability specification. The third is to create just enough feedback, because usability of usable security demands that designers and developers eliminate clutter and complexity for users on operation but we cannot predict every situation. At last of course the cost should never be forgotten; the user always strikes a balance between practicability and price, high cost of operation in the long run must be unacceptable.

5.6 Making Usable Security Acceptable to All

In order to achieve its intended purposes for the security in the real world universally, the security technology we worked so hard to bring to reality has to be enabled and applied everywhere it needs to be applied. When all the technical components are in place, usable security is also about making security acceptable to all stakeholders.

5.6.1 Solid Security Policy

No one can practice and enforce security without a solid security policy, even if the network is as small as a personal network consisting of only a handful of devices. Without such a security policy, the user may not know whether something bad has happened to his network. Good security always starts with a set of security policies, a policy that is based on risk vulnerability analysis and risk analysis. The preparation of a security policy begins with the preparation of creating usage policy statements, conducting risk analysis involving all stakeholders, and using risk analysis and development methods that integrate security and usability in specification. Although a user or a group of users (e.g., a family) does not have the structure of a large corporation or may not understand all the risks and their implications, it may be necessary to assign each resource into one of the three risk levels:

- low risk
- medium risk
- high risk

5.6.2 Universally Recognizable User Interface

There should be possibilities to recognize and universally accept security alerts, signals, symbols, etc. via a standardized user interface. The red–yellow–green traffic lights are one example, and the users know exactly what to expect when they come upon such a traffic light. For instance, when a driver sees a red light, he is expected to stop. When he does not, all kinds of consequences can occur as a result. Other good examples of well-known and well-established warning signs are the radiation sign and the skull sign to warn of potential danger. The user interface on a security device should be exactly like a set of traffic signs and regulations. For example, a signal on the device (a long tone or a long beep sound followed by two short sounds or a red or green LED blink once followed by twice) could indicate to the user that the device, which he is about to make a bank transfer authorization, is indeed secure.

5.6.3 Transparency

Transparency is big factor in applying the maximum amount of security for the users without intruding on their day-to-day routines. It should also minimize the physical and mental workload of users by limiting the range of simple actions and leaving complex actions to more tech-savvy users. Better yet, if complex actions are not required, leaving them out would not be such a bad idea and it also helps to minimize the choices and decisions for the user to no more than necessary. Interoperability, context-awareness, low-workload personalization, and simplicity all contribute to minimizing the choices and decisions for the user in a positive way.

5.6.4 Minimize Physical and Mental Workload of Users

When it comes to using security, it should work transparently and universally just like a classic and yet ubiquitous POTS (plain old telephone service) such that when a user picks up a telephone, he can dial whomever he chooses to without having to understand a thing how a telephone works, how it is connected to the network, or how and where the dial tone comes from and expect the call to be connected. Usable security should be designed the same way: extremely low complexity and requires a limited range of simple actions from the user to almost no action from the user other than picking up the phone and start dialing. In each case, the user makes no more decision than necessary. In fact, the only decision is whom should he dial. Security should help the users, and not impose upon or intrude into the users, nor should it make the life of the users difficult.

Of course, security is not mandatory for everyone and everything because not all communication and information are considered top secret. One-size security solution does not fit all users. Casual conversation between two teenage boys about a football match they have just watched should not be treated in the same secure level as a bank transfer authorization between a user and his bank. Different levels of security are required for different applications such that low-risk transactions do not require high level of security and they do not need to be over-secured. Over-security not only creates complexity, but also creates confusion. Different applications may require different devices. An elderly with limited mobility who is being cared for in a care facility may not require the same level of sophisticated device as a manager of a bank. The elderly may not want to learn about the risks and countermeasures, and since voice applications are predominantly the application of choice for them, a simple low-risk voice device that looks and functions much like a traditional POTS device may be sufficient, whereas the bank manager may require a sophisticated multimedia device with Internet access and with a higher level of security.

5.6.5 Seamless Transition

When users roam between two networks, the same level of security should be maintained across different devices and different platforms without user intervention. Seamless transitions make this interoperability across components and entities easy and secure. Standards can help define such a set of rules and requirements and interfaces for different devices across different platforms to interoperate and therefore maintain the same level of security across.

One of the biggest questions is who is responsible for the total control of the security. Obviously, if users are in total control, that will increase their overall workload. The users with an active role certainly will manage their security better, but only after a full understanding of what is involved and what it takes to be an active manager. To help the users decide what the best is for them, the service provider should be able to come up with a set of survey questions and recommend a policy for a particular user through a user-friendly interface. The following are examples of some of the questions:

- What is the level of security?
- What is the most important thing you want to protect?

Without fully understanding the user requirements, it is unlikely that a service provider would be able to provide a recommendation to the users, but at least the users can always rely on the network or service providers to provide domain-based security, which is becoming the de facto standard model for the service industry.

5.6.6 Rule of Trust

In managing risk, it may not always be possible to eliminate all risks. Therefore, there has to be a balanced role of trust for economics and human relationships in which certain risks are accepted and in some cases shifted to other parties. Shift in risk between parties needs to be made clear. It is important to raise the security awareness of the user and yet not overwhelm him with lots of details and responsibilities because chances are that in the security team the user assembly consists of a one-man team with the full assurance from his service provider's support in the event that the user may not be able to or may not have the expertise to completely manage his security services. Service providers may even choose to offer, through value-added services, user's network and resources to the manager and provide domain-based or network-based security simply because the network usually manages security better than individual users. Bear in mind that it is impossible to completely secure or eliminate all risks, but the goal is to create security awareness, minimize risks, and maximize the use of technology as much as possible by making the security usable, seamless, and transparent.

5.6.7 Human Element in Delivering Security

While it may not be possible for the user to do everything, such as a complete configuration, it is human nature to entrust some of his responsibilities to some higher "authority". This is the human element in delivering security, and it could mean someone acting as a trusted helper or agents who can provide a service for setting up security for the whole community because it is more likely that the users trust humans than technology. In addition, the trusted certificate authority concept had worked well for a number of years, ensuring that a person's public keys can be certified. After all, who wants to trust self-signed certificates? In a sense, most users are not security experts. They would have to rely on either humans or technology to make at least some of their decisions for them. Again, when it comes to security and technology, there are no one-size fits that fit all. Instead, everything should be tailored to the specific needs of the user.

5.6.8 Value-Based Security

Value-based security takes into account what is important to users but has to be done in a manner without creating too much work or personalization or customization. With security being an enabler, we offer a variety of new additional security and other services that will inevitably add to the value that the consumers perceive to receive and in turn add values to the provider's services as well. For instance, security may be offered as value-added services in tiers:

- Tier 1 with no security
- Tier 2 with selective security
- Tier 3 with premium or full-blown security that can even make the CIA envy.

5.6.9 Security Goes Beyond Authentication and Encryption

Most people think of security in terms of authentication and encryption alone, but reality is that making usable security acceptable to all goes beyond that. It is also about prevention in approving security changes and monitoring security of the network. When violations are detected, a quick response and restoration of affected services and systems are immediately warranted. Any holes through review after the fact should be plugged to prevent reoccurrence. A good security provides defense in depth, augmented by layers of security, people defense, security technologies defense, and operations defense. Better-informed users are better users. Better education means better awareness, which in turn means better users. We should always avoid stick-on or add-on security. This kind of security not only is complex, but proves low usability and high risks. After all these years, the lesson for all of us is that patches do not work all the time and costs more in the long run than to start system designing with security in mind in the initial stages.

5.7 Cost of Security

Security as a service is not free, and it may not be entirely clear who are the service providers and network providers in the traditional sense. One thing is almost certain that there is a clear division of responsibilities among the network service providers and end users, the former are responsible for network components and the latter for end terminals.

There is a cost associated with everything. Every one of the stakeholders has a component that adds to the cost of security, even with non-users such as attackers and hackers. Attackers and hackers incur a cost because security is put in place so that they can be excluded from receiving services. The end users who use beyond 3G services and applications should pay for the security they demand. Developers incur a cost because they spend much energy and time to design the system and applications with cost-effective security. System administrators and network operators who make their best effort to make sure the security of the network access and services also take the cost of security.

Because of the economic factor, it changes the landscape of accepting risks. We should take some balance between the function of security and the cost of security. The users always think much of the ratio of capability to cost. If the cost of the security is so high that few users can afford it, then the security is not usable and good enough.

It is possible to create applications and services for the future that are secure and usable. However, absolutely security cannot be measured objectively nor is it the right objective to aim for when seeking an overall optimum solution. Furthermore, no one builds services and applications without factoring the cost of building and providing such services and applications for B3G.

5.8 Conclusion and Prospect

We are accustomed to divide certain system architecture into several layers, each of them performs a special function and serves for the upper layer, for example, computer architecture is divided into seven layers. We must perform risk analysis on the "layer cake" architecture proposed, and a selection of example services.

Is it possible to create a system that is both absolutely secure and usable? Probably not, at least not at the present. Absolute security always indicates complexity and confusion, and at the same time high cost. Absolute security is not always the right objective to aim for when seeking an overall optimum solution.

Usable security should be considered as part of Total Cost of Ownership. In a typical system, the licensing costs are dwarfed by the Total Cost of Ownership, which includes training, maintenance, and interoperability while constraining adaptability. A truly usable system would have a large impact on the Total Cost of Ownership. It is especially common to externalize security costs by dumping them on the user. We need to strike a balance between perfect security and usable security, bearing in mind the economic factor. Security is driven by economy. Perfect security is not necessary for everyone or everything. Researchers around the world have been working hard in this area to provide or to add better usable security to everyday applications and services.

Some research centers are actively working on self-securing VOIP technology with key imprints to authenticate the call. Users are not required to be network experts. A world famous PC supplier is working on adding security to videoconference systems. Another well-known company is applying human–computer interaction (HCI) techniques to security functionality and creating principles of usable secure systems. Various other companies and academic institutions have contributed to cognometric authentication and value-sensitive design. There is even an annual symposium dedicated to usable privacy and security: Symposium on Usable Privacy and Security (SOUPS).

All of the researches will one day make it easier to build services and applications with better usability and security in mind from the beginning. Every system has a user. Better users make better systems. After all, a system is as good as the users who use it and it is as secure as the weakest link in the system. A certain person does not determine system security; the other way around, each stakeholder is involved in the overall security level. Different stakeholders should take on corresponding responsibility. For instance, designers should design the system architecture in a more concise, more effective, and more secure manner. Developers should develop a standard and convenient interface. As considerable common end users, we had better learn some security knowledge and instantly report the feedback of security threats and attacks.

And finally only when all of the components are solidly in place, we will have a system that can provide services and applications in a manner that is considered truly usable and secure.

References

[1] Cornwell, J., I. Fette, G. Hsieh, M. Prabaker, J. Rao, K. Tang, K. Vaniea, L. Bauer, L. Cranor, J. Hong, B. McLaren, M. Reiter, and N. Sadeh. "User-Controllable Security and Privacy for Pervasive Computing." In *Eighth IEEE Workshop on Mobile Computing Systems and Applications (HotMobile)*, February 2007.

[2] Josang, A., B. Alfayyadh, T. Grandison, M. AlZomai, and J. McNamara. Security Usability Principles for Vulnerability Analysis and Risk Assessment. *Proceedings of the 23rd Annual Computer Security Applications Conference*. Miami Beach, Florida. December, 2007.

[3] Li, Y., J. Hong, and J. Landay. "Topiary: A Tool for Prototyping Location-Enhanced Applications." In *Proceedings of ACM Symposium on User Interface Software and Technology (UIST2004), CHI Letters* 6, no. 2 (2004): 2004

[4] Zurko, M.E. User-Centered Security: Stepping Up To The Grand Challenge, Annual Computer Security Applications Conference (ACSAC), December 2005.

6

Conclusion

L. Sørensen

CMI, Aalborg University, Aalborg, Denmark

The purpose of this book was to present different perspectives on elicitation of user requirements for new wireless services. Four different cases, which seemingly have a relatively straightforward and similar software engineering process, however, pose a significant number of variations and challenges in the user requirements elicitation process. In this section, the different book contributions are analyzed in terms of what we can learn about the user requirements elicitation process in practice.

The cases, presented, all focus on different types of users, in terms of age, context in which the application/service is to be used, and who represents the user. The cases particularly focus on the following elements: the users and the context, the involvement of users, user requirements and whom they represent, the development process, and other areas. Each of these is discussed in the following.

6.1 The Users and the Context

In several of the book chapters, the need for understanding the users, who they are, and what they need is essential for the development of the service or application. In the "Developing Mobile Application for Children" chapter, it is clear that understanding the children, the different children age groups, and the focus of the school was essential. The project goals were formulated based on this understanding ending up with goals such as inclusion, education, and technical development. It was the understanding of these elements that led to how the application was developed in terms of musical theme and screens based on the children's drawings.

In the chapter on "Developing Applications for Construction Workers", focus was on understanding the particular context in which the users (construction workers) were operating. Also in this chapter, it was fundamental to understand the users, how savvy they were to use mobile phones and how they work and operate in an outside and weather changing environment. The selection of technology (devices), and the graphical interface design were adapted to this knowledge and resulted in a far more simple and sound/vibration led application that was originally envisioned.

In the chapter on the FarmerNet application, we here hear about a third group of users, namely relatively poor farmers in Sri Lanka. The understanding of the users with barriers of illiteracy, and low technology exposure, and the low income were essential parameters. But also the understanding of how farmers traditionally sold their trading to middlemen was important and was essential for the requirement specification, since that created an understanding of what the users usually did in selling their produce.

In the "Conflicting Interests in User Requirements..." chapter, the lack of knowing the user caused the development of a failed service that the user not had any interest in. The way that stakeholders were formulating the user requirements was representing political goals more than the users' needs. Therefore, in this case, the author (Sørensen, J.) questions whether this can be called user requirements.

In the literature, it is clear that the understanding of who the users are is essential for successful user requirements elicitation. In, for example, Porras et al. (2014), it is shown that different users and ages have differing and changing user requirements, which demands a special focus on the particular group of users in the requirements elicitation process. Other authors also support this reflection. See for example Ståhlbröst (2008) or Edvardssoon et al. (2006) where different studies point toward the same conclusion. Furthermore, the study from Frohlich et al. (2009) describes how studies have been made in a village in India with the use of digital storytelling for requirements elicitation with an inhomogeneous user group and limited technology options.

6.2 The Involvement of Users

The three cases on the application development for children, construction workers, and the farmers are examples of how different the user involvement can take place. In all cases, the users are involved in different ways according to what and how they can be expected to be contributing to the definition of user requirements. In the cases, the use of low-fidelity prototypes was important.

Many different approaches were used to involve the users in the three projects: interviews, monitoring, focus groups, sketching (paper and canvas), role-play, and many more. What can be said to be essential is that the user involvement approach needs to be adapted to the users and the situation they are in. This again calls for a note on the work by Frohlich et al (2009) where citizens of an Indian village were involved in mobile digital storytelling for elicitation of design requirements. This study illustrates that users must be involved in a development process with a clear focus to the context, user group, the IT knowledge, and availability, but also with an understanding of what interests (and daily needs) that the users have much more than the interests of the software developers.

6.3 User Requirements and Whom They Represent

In all of the book's cases, it is mentioned that the user requirements originate from a combination of project group/stakeholder formulated requirements as well as from the real users. Again in the three cases involving children, construction workers, and farmers, there was set up a process in which the different user requirements were validated and changed according to new understanding of the users' needs—which secured a good adaptation and mix of the user requirements.

In the case on the personalized media service, this process was not in force. Here, different stakeholders in the media organization formulated the user requirements, and decision-makers again changed them out of a self-interest (politically or technically). As it is discussed in the case, the question is whether this can be characterized as user requirements or not. This case clearly shows that in real life, the user requirements elicitation process can be a process of negotiating and shaping a new service driven by an internal stakeholder and decision-making process in the organization. This process could be characterized as a traditional development process in which the users are not involved. The case illustrates in full the challenges that lie in this.

6.4 The Development Process

The case with the farmers in Sri Lanka is the case which was mostly controlled by an innovation ideology and application of the star gate process to manage the software development process. The case describes that this was a good approach and there are not clear observations of problems in this. This approach resulted in a good correlation between focusing on the farmers but also at the external environment, which was important for the sake of the farmers if they were to use the application.

In the case with the children, there is mentioned that a central management for integration knowledge was not established. The case describes how that ambition levels were set un-realistically high and not with a clear understanding of the technology and its challenges. The programming of the application was not done in a sufficiently transparent way leaving many decisions on the programmers' hands alone. The case describes that the development process was characterized by knowledge and information sharing gaps between different groups in the project and that it caused a lack of information and decision-making flow.

This case clearly illustrates some of the challenges of practical application and service development, namely the gabs within the project where different scientific groups cannot communicate well (either because of different languages and understanding of the language or because of the lack of a management form which secured a smooth knowledge and decision-making process).

6.5 Other Areas

One chapter in the book touches upon an area not discussed in the other chapters, namely security as a part of any service and application development. This is referred to as usable security. This chapter sets focus on that security must be addressed as a broad concept, and to raise the level of understanding and use of security functionalities, it should be implemented in a transparent and user-friendly way. This chapter advocates for that security must be placed in any application or service in such a way, so the user can control, understand, and use it correctly. As it is now, it is one of the elements in software, which are often disregarded or handled in a technical way more than a transparent and user-friendly way. Looking at usable security, as it is suggested, users would easily be able to reflect and provide feedback to this element as well in a service and application development process. This would take the concept from being a purely technical-oriented element to being part of the user requirements elicitation process.

6.6 Final Comments

This book illustrates some of the practical challenges in elicitation of user requirements and in developing software for a particular group of users. The book does not in any way claim to be neither general nor sufficiently statistically representative to be able to conclude in general terms. In spite of this, the book would like to point toward a number of final guidelines that represent the cases and learning of these. These are as follows:

- Study the user for which the service or application is to be used by. The study of the user is essential for setting goals for the software development.
- Study also the external environment where the user will be using the service/application. The external environment shall be understood as the working environment (as for example of construction workers in Finland) but also as an understanding of the communication process which the service or application must be part of. By this, it is meant that the service or application is defined also by external communication links and that this must be addressed from the beginning (the example from this book is the FarmerNet which should be used by local farmers but completely change the way that farmers sell their products). It is important to understand how many changes the service or application will place upon the existing user behavior.
- Involve users in software development—and involve the real users in the way they can be involved best (dependent on age, context, education level, interest and natural needs, etc.).
- Design for the users and address gender, age, context, income, affordability, etc.
- Place focus on the organization of the user requirements elicitation process and the rest of the service/application development process. The requirements elicitation process must be clearly linked to all other elements of the project. It demands a management structure that addresses this element.
- Be aware that user requirements not necessarily are representing users, but they are results of political and decision-making processes of internal stakeholders in an organization. It is still user requirements, by definition, but is hard to compare to user requirements which have been elicited and verified by real users.
- Adapt the concept of usable security to allow that security becomes a central element of any user requirement elicitation process.

References

[1] Edvardssoon, B., A. Gustafsson, P. Kristensson, P. Magnusson, and J. Matthing, *Involving Customers in New Service Development*. London: Imperial College Press, 2006.

[2] Frohlich, D.M., D. Rachovides, K. Riga, R. Bhat, M. Frank, E. Edirisinghe, D. Wickramanayaka, M. Jones, and W. Harwood. *StoryBank: Mobile Digital Storytelling in a Development Context.* CHI 2009–Mobile Applications for the Developing World, April 8th, 2009, Boston, MA, USA, 2009.

[3] Porras, J. et al. *The User 2020–Vision. World Wireless Research Forum Outlook.* 2014. (To be published).

[4] Ståhlbröst, A. *Forming Future IT. The Living Lab Way of User Involvement.* Doctoral thesis at Luleå University of Technology, Sweden, 2008.

Index

About the Editors

Lene Tolstrup Sørensen is Associate Professor at Center for Communication, Media and Information Technologies, Department of Electronic Systems, Aalborg University Copenhagen. She holds a Master in Engineering (1990) and Ph.D. (1994) from the Technical University of Denmark. She has worked professionally in the areas of IT strategy, interaction design, and requirements engineering for more than 20 years. She has participated in a number of research projects financed by the European Union, the Danish Research Councils, and the Danish Technology Research Council. She has published nationally and internationally more than academic publications in international journals, books, and conference proceedings. The main research areas are as follows: user requirements engineering, user involvement in mobile and wireless technology developments, and design of user centric services.

Knud Erik Skouby is professor and director of Center for Communication, Media and Information technologies, Aalborg University Copenhagen. He has a career as a university teacher and within consultancy since 1972. His working areas are as follows: *Techno-economic Analyses; Development of mobile/wireless applications and services: Regulation of telecommunications.*

He is project manager and partner in a number of international, European, and Danish research projects. He served on a number of public committees within telecom, IT, and broadcasting and served as a member of boards of professional societies, as a member of organizing boards and evaluation committees, and as invited speaker on international conferences; he published a number of Danish and international articles, books, and conference proceedings. He is board member of the Danish Independent Research Council and the Danish Media Committee. He is the chair of WGA in Wireless World Research Forum, Special Advisor to GISFI, and department chair of IEEE Denmark.
E-mail: skouby@cmi.aau.dk

About the Authors

Jing Chen is a member of the senior research team in the wireless security group in Huawei's Shanghai Research Center. He received Master in Xi'dian University, Xi'an, China. He has worked in wireless security area since the early 2000s. His current research interests include security of wireless system, IP security, trusted computing, etc. He has extensive involvement in various security standardization activities in 3GPP, IETF, and CCSA. He currently is the serving vice-chairman of WG2 (Wireless Security Working group) within TC8 of CCSA. Jing also contributed to numerous publications in some leading journals such as the Journal of Cyber Security and Mobility. In addition, he has more than 20 patents granted and/or pending.
E-mail: eric.chenjing@huawei.com

Philip Edge, PhD, is an Associate Consultant at eNovation4D. Philip trained originally as a biologist and subsequently carried out research in the history of medicine. This was followed by 25 years of working in scholarly publishing, during which he was involved in the early years of developing scholarly information for delivery on the Internet. Since 2002, he has worked both on the digital management of research information in developing countries, and in the ICT4D sector, addressing the challenges and opportunities of the digital information chain which joins the researcher and the end user. Philip has carried out work for the Food and Agriculture Organization of the UN, The World Bank, and other international organizations.
E-mail: edgedr@gmail.com

Kari Heikkinen Dr. Sc (Tech) is Associate Professor of User-Centric Software Engineering at the Lappeenranta University of Technology, Finland. He received the Dr. Sc. (Tech.) degree from the Lappeenranta University of Technology, Finland, in 2005 about Conceptualization of User-Centric Personalization Management. His work and interests include collaborative learning, learning processes, learning applications, and practice-based innovation.

He has published over 65 scientific articles in his expertise areas. He has 8-year experience in project management in both national and EU projects. E-mail: kari.heikkinen@lut.fi

Harsha Liyanage, PhD, MBA, Lead Consultant eNovation4D is an ICT4D sector expert with over 18 years of experience in a variety of industries including wireless, telecenters, start-up social enterprises, multi-stakeholder partnerships, donor relations, social impact assessment, project design, and economic sustainability.

Harsha is the Founder of Sarvodaya-Fusion, the ICT4D program of Sarvodaya, Sri Lanka. His long involvement with the grassroots sector, both as Deputy Executive Director of Sarvodaya-Fusion (from 1994 to 2004) and as Honorary Vice-President of Sarvodaya, has led to consultancy work for many organizations including the UN, IDRC, and other international development organizations. In 2010, Harsha founded eNovation4D Ltd., a UK-based consultancy company with the objective of disseminating ICT4D expertise between the Global South and Global West. With a vision of 'Fixing the "D" of ICT4D,' his work continues in countries in Africa and South Asia. E-mail: harsha@enovation4d.co.uk

Pirkko Paananen-Vitikka (PhD) is Adjunct Professor of Music Education in University of Oulu, and Head of Music School in North Eastern Lapland, Finland. During her academic career, she has worked as researcher, senior researcher, and special researcher in music education at the Department of Music, University of Jyväskylä, Finland, as well as senior assistant of music education at the Department of Teacher Training, University of Jyväskylä, Finland. She has also worked as music teacher and piano teacher. Her research areas are children's musical development and creativity, special education, educational technology, and new learning environments of music. She has participated as a researcher in charge in mobile music technological projects, such as the MobiKid 2005–2007 pilot study (www.jyu.fi/hum/laitokset/musiikki/en/research/development/project/mobilemusic/) and EU FP7 UMSIC 2008–2011 (www.umsic.org). In these projects, she has focused in child-centered, game-based pedagogical and musical design of the software for young and school-aged children as well as testing the impacts of these new learning environments. Her hobbies include music composition and song writing. E-mail: pirkko.a.paananen@jyu.fi

Jari Porras Dr. Sc. (Tech) is Professor of Distributed Systems and Wireless Communications and currently head of the Software Engineering and Information Management Department at the Lappeenranta University of Technology, Finland. Prof. Porras received the Dr. Sc. (Tech.) degree from the Lappeenranta University of Technology, Finland, in 1998 about modeling and simulation of communication networks in distributed computing environment. His work and interests include wireless networks, ad hoc networking, peer-to-peer computing, and aspects of cloud computing as well as distributed computing and distributed environments, simulation and modeling. He has published over 100 scientific articles in his expertise areas. Prof. Porras acts as the chair of the Working Group B–Services, Devices and Service Architectures in the Wireless World Research Forum (WWRF). Currently, he is responsible person in LUT for the European Erasmus Mundus PERCCOM programme.
E-mail: trappis@lut.fi

Ross Purves is a Lecturer in Education Studies at De Montfort University in Leicester, UK. He studied music at City University and the Guildhall School of Music and Drama, before completing an MA in music education at the Institute of Education, University of London, and a PGCE in post-compulsory education at the University of Greenwich. After graduating, Ross worked as a computer development officer in the Department of Computer Science, University of Exeter, and subsequently as an education research officer at the University of Roehampton and the Institute of Education. Between 2009 and 2011, he was a research officer on the European Commission's Seventh Framework-funded 'Usability of Music for Social Inclusion of Children Project'. Ross was also Joint Course Manager for Music at Luton Sixth Form College for six years and Subject Coordinator for the Music on a large, secondary school-based initial teacher education program. Publications embracing musicians' and teachers' professional development, music technology education, and aspects of human–computer interaction have appeared in journals and edited international handbooks including the British Journal of Educational Research, the British Journal of Educational Psychology, Issues in Technology and Teacher Education and the Bulletin of the Council for Research in Music Education, and the 2012 OUP International Handbook of Music Education. Ross has presented research at various European education conferences and is an experienced performing musician and arranger.
E-mail: ross.purves@dmu.ac.uk.

Jannick Kirk Sørensen is Assistant professor at Center for Communication, Media and Information technologies (CMI) at Aalborg University, Denmark. His research fields are interaction design, service innovation, personalization, and user involvement, particularly in the field of media technology. He holds a PhD degree from University of Southern Denmark (2011) on the topic of personalized webpages for public service broadcasters. Prior to his academic career, he worked in the media industry as a journalist, interaction designer, and sound engineer.
E-mail: js@cmi.aau.dk

Professor Graham Welch holds the Institute of Education, University of London Established Chair of Music Education (since 2001). He is Immediate Past President of the International Society for Music Education (ISME) (from 2012 to 2014), elected chair of the internationally based Society for Education, Music and Psychology Research (SEMPRE) and past co-chair of the ISME Research Commission of ISME. He holds Visiting Professorships at universities in the UK and overseas, and he is also a member of the UK Arts and Humanities Research Council (AHRC) Review College for music. Internationally, he has acted as a specialist external consultant on aspects of children's singing and vocal development for UK and Italian Government agencies, as well as for specialist research centers in the USA, Australia, and Sweden. He published more than 300 research articles.
E-mail: g.welch@ioe.ac.uk

Marcus Wong received his Master of Arts Degree in Computer Science from Queens College of City University of New York (USA). He has over 20 years of experience in the wireless network security field with AT&T Bell Laboratories, AT&T Laboratories, Lucent Technologies, and Samsung's Advanced Institute of Technology. He holds Certification of Information System Security Professional (CISSP) from the prestigious International Information Systems Security Certification Consortium (ISC2).

Marcus has concentrated his research and work in many aspects of security in wireless communication systems, including 2G/3G/4G mobile networks, personal area networks, and satellite communication systems. Marcus joined Huawei Technologies (USA) in 2007 and continued his focus on research and standardization in 3GPP, WiMAX Forum, IEEE, and IETF security areas. As an active contributor in the Wireless World Research Forum (WWRF), he has shared his security research on a variety of projects contributing toward whitepapers, book chapters, and speaking engagements.

In the past, Marcus has held elected official positions in both WWRF and 3GPP, serving as the Vice-Chairman of WWRF Working Group 7 from 2007 to 2012 and as the Vice-Chairman of 3GPP SA3 from 2009 to 2011. He also served as guest editor in the IEEE Vehicular Technology magazine. He is an author of the book "Femtocells: Secure Communication and Networking" along with a number of journal papers and whitepapers in leading publications, including that of the Journal of Cyber Security and Mobility. In addition, he has numerous patents granted and/or pending.
E-mail: mwong@huawei.com

Lijia Zhang received her M.S. degree in Telecommunication from Beijing Jiaotong University in China. After completing her education, she joined Huawei's Research Center in Beijing. Currently, she is a senior security researcher focusing on many aspects of security research. She has worked in the areas of wireless security and IP security. She is actively involved in security standardization activities and has contributed a large numbers of proposals to both 3GPP and IETF standardization bodies. In addition, she has more than ten patents granted and/or pending.
E-mail: emma.zhanglijia@huawei.com

Lightning Source UK Ltd.
Milton Keynes UK
UKOW06n1146270117

293027UK00002B/5/P